Ulrich Beck

Ulrich Beck

A Critical Introduction to the Risk Society

Gabe Mythen

Pluto Press
LONDON • STERLING, VIRGINIA

First published 2004 by
Pluto Press
345 Archway Road, London N6 5AA
and 22883 Quicksilver Drive, Sterling, VA 20166–2012, USA

www.plutobooks.com

British Library Cataloguing in Publication Data
A catalogue record for this book is available from the British Library

ISBN   0 7453 1815 0 hardback
ISBN   0 7453 1814 2 paperback

Library of Congress Cataloging in Publication Data applied for

10   9   8   7   6   5   4   3   2   1

Designed and produced for Pluto Press by
Chase Publishing Services, Fortescue, Sidmouth, EX10 9QG, England
Typeset from disk by Stanford DTP Services, Northampton, England
Printed and bound in the European Union by
Antony Rowe Ltd, Chippenham and Eastbourne, England

Contents

Acknowledgements

A book cannot be written alone and there are a number of people whose knowledge, vitality and generosity have enabled me to produce this piece of work. I am grateful to Con Lodziak, Steve Taylor and John Tomlinson, each of whom sparked in different ways my interest in social theory and have acted as inspirational mentors. Anne Beech helped push the idea through in the early stages and provided invaluable editorial guidance throughout. The book has benefited from exchanges with colleagues at Manchester Metropolitan University, in particular Mark Banks, Helen Jones, Bernard Leach, Liz Marr, Katie Milestone, Jenny Ryan, Shirley Tate and Sandra Walklate. Thanks are also due to the following people, whose ideas, prompts and friendly criticisms helped knock off the rougher edges: Finn Bowring, Simon Cross, Simon French, Luke Goode, Linda Janes, Phil Kelly, John Maule, Michael Mehta, Kate Millar, Anne Nisbet, Richard Ronald, Jeremy Tatman and Corinne Wales. Finally, I am indebted to my family and my partner for offering the unstinting support and encouragement which sustained me throughout. You know who you are.

Introduction

In western cultures, the latter half of the twentieth century has been described as an epoch of flux, uncertainty and rapid social change (Bauman, 1991; Marwick, 1990). During this period, distinct transformations in the structure and functions of dominant institutions generated a complex mix of liberties and constraints (Giddens, 1991; 1994; Waters, 1995). Since the Second World War, the building blocks of society have effectively been shaken up and relaid. Far-reaching transformations in family structure, employment patterns and welfare provision have redrawn class boundaries, shuffled gender roles and chopped up social identities (Beck, 2000a; Hughes and Fergusson, 2000). As the twenty-first century unfolds, the process of globalisation continues to disperse through economies and political institutions, rendering visible the connections between global shifts and local actions (Robertson, 1992, Tomlinson, 1999). Economic convergence, political fluctuation and national insecurity have become the motifs of the age. We are living in a 'runaway world' stippled by ominous dangers, military conflicts and environmental hazards. As a result, increasing portions of our everyday lives are spent negotiating change, dealing with uncertainty and assessing the personal impacts of situations that appear to be out of our control. In one way or another, the defining markers of modern society are all associated with the phenomenon of risk. In contemporary culture, risk has become something of an omnipresent issue, casting its spectre over a wide range of practices and experiences (Adam and van Loon, 2000: 2; Lupton, 1999a: 14). Locally, risk emerges as a routine feature of existence in areas as diverse as health, parenting, crime, employment and transport. Globally, concerns about air pollution, the state of the world economy and the spread of Acquired Immune Deficiency Syndrome (Aids) are all underscored by risk.

As if to accentuate the instability of the modern era, the events of September the eleventh have acted as a high-voltage shock to the capitalist system. Post 9/11, something fundamental has changed in the way we perceive the concepts of safety and danger. Following a backdraft of concern about bioterrorism, twitchy politicians have advised citizens to stock up with essential foodstuffs and bottled water. On an international stage, world leaders talk about the menaces of living in a 'post-secure' world in which an 'axis of evil' threatens to spread 'global terror'. In the words of the British Prime Minister, 'September the eleventh was not an isolated event, but a tragic prologue ... our new world rests on order. The danger is disorder. And in today's world it can now spread like contagion.'[1] Putting aside the political rhetoric, 9/11 has acted as a long overdue wake-up call for inhabitants of the affluent western world. The tragic incidents in New York and Washington illustrate that the unthinkable can and should be thought. After all, it has already happened.

Since the attacks on the World Trade Center and the Pentagon, security investigations in several European countries have uncovered worrying evidence of intent to use biological and chemical substances amongst terrorist groups. Dirty bombs, anthrax, ricin and sarin have crept into the public vocabulary. The hyper-uncertain climate which has taken hold post 9/11 speeds a clear idea of what it means to live in a 'risk society'. Layered over the top of longstanding everyday hazards, current anxieties about the threat of terrorism have added to a general feeling of public unease. As Jasanoff comments: 'Just as a century ago, the idea of "progress" helped to name an optimistic era, so today "risk", by its very pervasiveness, seems to be the defining marker of our own less sanguine historical moment' (Jasanoff, 1999: 136).

Despite its ubiquity, the meaning of risk remains indeterminate. In contemporary society, the effects of various risks are keenly contested by politicians, scientific experts, media professionals and the general public. It is this very lack of consensus that makes risk such a fascinating topic of inquiry; and one which is always likely to produce disagreements. In western cultures, the meaning of risk has evolved alongside the development of social institutions, the economy and the welfare state. Following on from the Enlightenment period, the rapid expansion of scientific, technological and medical knowledge created an assemblage of expert systems of risk calculation, assessment and management. Social commentators of different persuasions are in consensus that the application of various forms of institutionalised knowledge about risk has enabled western cultures to eliminate a succession of threats to public health that blighted earlier epochs (see Furedi, 1997; Giddens, 1991: 116; North, 1997).[2] Accordingly, the incidence of infectious and epidemic diseases has fallen dramatically over the last 150 years (Smith, 2001: 148). Due to the capacity of science and medicine to improve both life expectancy and quality of life, various forms of risk regulation have become enshrined in health, medicine, law and government. A proliferation of technical and scientific knowledge about risk and the dissemination of regulatory procedures have undoubtedly fostered more acute forms of public consciousness. In the late modern period, public awareness of risk has also been influenced by the extension of the mass media and the growth of new information and communication technologies. The fluidity of information has enhanced channels of public communication and is propagating more visible debates between stakeholders (Strydom, 2003: 2).

However, as far as social knowledge goes, developments in technology, medicine and science have produced something of a cleft stick. Greater access to information about risk has empowered people to enact positive lifestyle changes, particularly in relation to health, fitness and diet (Beck-Gernsheim, 2000; Lupton, 1999b: 62). Yet the expansion of information has also caused conflicts over the meaning and impacts of risk amongst competing interest groups. Despite its enjoying comparative health and longevity, transboundary dangers cast a shadow of discomfiture over contemporary western society. Thus, the implicit bargain for techno-scientific development and heightened risk consciousness might well be the amplification of insecurity:

> Over the past months and years we have endured the SARS crisis, the BSE scandal and the foot-and-mouth epidemic. We've been warned of deep-vein thrombosis from air travel, brain cancer from mobile phone radiation and mutations from genetically modified organisms. We've been told that climate change threatens our coastlines, antibiotic resistant viruses threaten our children and wayward asteroids threaten our planet. (Bird, 2003: 47)

At the level of risk perception, advancements in knowledge have failed to result in a more secure social climate. As the means of combating certain threats are promulgated, techno-scientific research generates more complex questions and issues. In matters of risk, it would seem that 'the more we know, the less we understand' (van Loon, 2000a: 173). This paradox enables us to appreciate why individuals in the West live comparatively longer and healthier lives, whilst simultaneously feeling less safe and secure (Pidgeon, 2000: 47; Sparks, 2003: 203).

In the last three decades, the availability of information about risk has been aided by the diffusion of media technologies. The broader circulation of risk communications within the mass media has undoubtedly enhanced awareness of risk and intensified public scrutiny of social institutions (Fox, 2000: 1; Wynne, 1996). The rising cultural profile of risk has also aroused more fundamental concerns about the relationship between individuals, institutions and society. In some instances, contestation and deliberation about risk have acted as a conduit for the articulation of broader ethical concerns (ESRC Report, 1999: 20; Vera-Sanso, 2000: 112). In addition to being construed as a scientific and economic affair, risk is also interpreted as a political and a moral issue (Caygill, 2000: 155). The debate currently taking place about the use of reproductive technologies for human cloning stands as a case in point.

The intensification of interest in risk amongst the media, politicians and the public has been mirrored by growing fascination with the subject within academia. Scholars of politics, science, health, economics, employment relations and the environment have all contributed to a colourful debate, giving rise to an ever-expanding number of research projects, study groups and university departments specialising in risk. However, although the language of risk is prolific, the concept itself remains cloaked in ambiguity. The residual lack of clarity surrounding both the constitution and the social impacts of risk have made it an irresistible area of inquiry for the social sciences. As a means of conceptualising risk, four paradigms have evolved within the social sciences. First, inspired by the pioneering work of Mary Douglas (1966, 1982, 1985, 1992), anthropological approaches have emerged. Anthropologists such as Douglas have investigated variations in understandings of risk between individuals and groups around the globe. Differences in risk perception have been unearthed and accounted for through particular patterns of social solidarity, world-views and cultural values. In recent times, the anthropological approach to risk has been revitalised by the efforts of Caplan (2000a), Bujra (2000) and Nugent (2000). Second, within the domain of social psychology, the psychometric paradigm has focussed on individual cognition of risk. In this oeuvre, Paul Slovic (1987, 1992, 2000) and his colleagues have developed psychometric methods of testing in order to determine which risks are perceived to be harmful by the public. Psychometric approaches have been oriented towards establishing the perceived constitution of various risks and the effects of this on estimations of harm. On the basis of psychological research, the heuristics and biases that commonly affect individual perceptions of risk have been delineated.

Third, the governmentality approach to risk has been fashioned by a crew of theorists deploying Michel Foucault's writings on the disciplinary effects of discourse (Foucault, 1978, 1991). In this spirit, theorists such as Castel (1991), O'Malley (2001) and Dean (1999) have accentuated the role of social institutions in constructing understandings of risk which restrict and regiment human behaviour. Fourth, the risk society perspective assembled by Beck (1992) and seconded by Giddens (1998, 1999) has demarcated the pervasive effects of risk on everyday life. Both Beck (1999: 112) and Giddens (1998: 28) maintain that the process of modernisation has spawned a unique collection of humanly produced risks. The deleterious consequences of these 'manufactured risks' span the globe, giving rise to radical changes in social structure, politics and cultural experience.[3] According to Beck, contemporary western cultures are party to a sweeping process of change, generated by the individualisation of experience and the changing logic of risk distribution (Mol and Spaargaren, 1993: 440). In the risk society narrative, seismic shifts in the relationship between the natural and the social necessitate refreshed ways of conceptualising society:

> A new kind of capitalism, a new kind of global order, a new kind of politics and law, a new kind of society and personal life are in the making which both separately and in context are clearly distinct from earlier phases of social evolution. Consequently a paradigm shift in both the social sciences and in politics is required. (Beck, 2000c: 81)

What is remarkable about our current situation is the extent to which the global and the local intertwine. Decisions made at global altitude – for, example, about international trading, nuclear power or global warming – produce knock on consequences for local activities. Similarly, local practices – overproduction, regional conflict or the production of poisonous emissions – generate consequences which impact in distant regions.

In the last decade, the risk society perspective has been hugely influential, serving as a stimulus for academic, environmental and political dialogue (see Caplan, 2000a: 2; Adam and van Loon, 2000: 1). Beck's extensively referenced *Risk Society: Towards a New Modernity* (1992) is considered to be a landmark text in social and cultural theory (McGuigan, 1999: 125; Reiner et al., 2003: 176).[4] Indeed, the term 'risk society' has become something of a lingua franca, capturing the imagination of the media and the public. Such currency – one hesitates to say notoriety – has only served to enhance Beck's status as a 'zeitgeist sociologist' (Skinner, 2000: 160). For Beck, the concept of risk unlocks and defines the essential characteristics of modernity. However, *Risk Society* (1992) not only delves into the muddy waters of risk, it also provides a reflection of the modern condition and a sweeping narrative of social reconfiguration. For sure, the risk society thesis is about much more than just risk. Beck's work also examines the broader interrelationship between humans and the environment, the effects of institutional change on social experience and the changing dynamics of politics.

What is more, the risk society thesis is also exceptional by way of academic method. As Bronner (1995: 67) notes, Beck has 'an extraordinarily lively style, a provocative way of raising questions, and a genuinely experimental sensibility'. The risk society perspective goes well beyond the parameters of the social sciences, borrowing from art, poetry and philosophy. Whilst Beck's unconventional style of writing has something of a postmodern quality, the content of the argument is indisputably modernist (Beck, 2002: 17; Dryzek, 1995). The risk society thesis is an attempt to capture the essence of social experience along the paths previously trodden by Marx, Weber and Habermas:

> What I suggest is a model for understanding our times, in a not unhopeful spirit. What others see as the development of a postmodern order, my argument interprets as a stage of radicalised modernity ... where most postmodern theorists are critical of grand narratives, general theory and humanity, I remain committed to all of these, but in a new sense ... my notion of reflexive modernity implies that we do not have *enough* reason. (Beck, 1998a: 20)

Although the modernist rationale underpinning the risk society thesis has been questioned (Bujra, 2000), there is little doubt that Beck's work has been instrumental in forcing risk onto the academic agenda. The risk society perspective has been pivotal in the evolution of cross-discipline debate between sociology, cultural studies, politics, geography and environmental studies. As a result, a medley of eclectic collections have mobilised Beck's theory of risk as a touchstone for broader discussion about the role of technology, health and politics in society (see Adam et al., 2000; Caplan, 2000a; Franklin, 1998). A further bunch of theorists have sought to examine specific strands of the risk society thesis, such as the portrayal of reflexivity, the functions of the media or the logic of political distribution (see Lash, 1994; Cottle, 1998; Scott, 2000). Finally, several academics have offered progressive reviews of the risk society perspective as it pertains to specific areas of interest, such as the environment, postmodern culture and the psychology of anxiety (see Goldblatt, 1995; McGuigan, 1999; Wilkinson, 2001).

Despite great excitement about its explanatory possibilities, a holistic analysis of the risk society perspective has not been forthcoming. Bearing in mind Beck's prolificacy and his academic status, the absence of a systematic deconstruction of the risk society perspective is nothing short of remarkable. In this book, I seek to fill the lacuna by following two interconnected seams of inquiry. Firstly, in an applied fashion, I recount the central tenets of the risk society perspective, testing its credibility in relation to existing theoretical and empirical evidence. Secondly, in a more thematic vein, the risk society thesis is employed as a vehicle for discussing the wider impacts and effects of risk on various social domains.[5] Insofar as the risk society theory will be compared and contrasted with contemporary cultural practices and experiences, our textual journey also makes use of anthropological, psychometric and governmentality approaches as tools for comparison and critique.[6]

From the outset, it is worth identifying those interests which lie on, or beyond, the margins of this inquiry. The book does not seek to provide a detailed historical review of the concept of risk. Although the social evolution of risk is more than an occasional theme in the following pages, I do not seek to provide an exhaustive examination of the history of risk within the social sciences. Nor is the book intended as either a straight biography, or a precis of Beck's academic work. Given the span and sophistication of Beck's writing – from cosmopolitanism to the nature of love – this task is happily left to others. To recapitulate, the discrete object of scrutiny is the risk society perspective as expounded by Beck at various stages of his academic career (1992, 1995a, 1998a, 1999, 2002). Hence, our mainstays of textual discussion will be Beck's most renowned works, *Risk Society* (1992), *Ecological Politics in an Age of Risk* (1995a) and *World Risk Society* (1999).[7] Nonetheless, the book is structured around exploration of issues, rather than sequential textual deconstruction and draws across the spectrum of Beck's writing, applying relevant texts to appropriate subject areas.

It is also worth logging the inevitable problems of interpretation that have been associated with Beck's work (see Smith et al, 1997: 170). A number of hermeneutic difficulties arise out of Beck's predilection for ambiguity, oscillation and dramatic effect (Goldblatt, 1995: 154). In the first instance, the big ideas are often played out on a highly abstract, theoretical plane (Cottle, 1998: 10). To further obfuscate matters, Beck is partial to switching between tenses within chapters.[8] On occasion, such a mixed style of communication makes it difficult to decipher whether one is reading about the past, the present or the future. Indeed, Beck often writes in what might be referred to as a hypothetical present tense, 'as if' the scenario recounted were actually taking place. In other passages, a series of future scenarios are offered up for consideration (Beck, 1992: 223–35; 1997: 90–4; 2000: 150–79). As we shall see, although this unorthodox mode of narration has undoubted projective benefits, the frequent variations of style do tend to come at the expense of theoretical clarity. In *Risk Society* (1992), Beck changes the style of delivery, scoots from point to point, leaves layers of ambiguity and wilfully changes his mind. Such an extraordinary format and style make it difficult to subject Beck's work to the usual methods of analytical scrutiny. I have tried, where possible, to tread a path which remains sensitive to the experimental nature of the risk society thesis, but does so without losing the cutting edge necessary for effective sociological criticism.

In its entirety, the book challenges the risk society thesis by exploring and reevaluating the relationship between risk, structural change and lived experience. I wish to construct a long overdue critique of the risk society thesis which refutes the claim that the dispersal of risk engenders a radically 'new mode of societalization' (Beck, 1992: 127). In contrast to the universalism inherent to the risk society perspective, I will be emphasising the complexity and multidimensionality of everyday negotiations of risk. In order to fashion this critique, I trace Beck's approach, outlining the alleged impacts of risk on vital social domains, such as politics, science, the environment and personal relationships. Whilst such descriptive shadowing is a necessary prerequisite for understanding, in each chapter, subsequent analyses track empirical and theoretical evidence in order to question, refine and extend the risk society thesis.

In Chapter 1, the concept of risk is introduced. At this juncture, the composition and functions of risk as a social, economic and cultural construct are unpacked. This rudimentary discussion is complemented by an account of the risk society thesis which identifies the 'pillars of risk' and 'icons of destruction' which support Beck's argument. In Chapter 2, general mapping of the risk society gives way to a more focussed consideration of the production of environmental risk. Here, the relationship between the natural and the social is bought into focus via an evaluation of the impacts of environmental risks on the ecosystem. Using a series of vignettes, the chapter probes the risk society narrative, fleshing out the material and ideological consequences of environmental despoliation. The third chapter unravels the role of dominant social institutions in building and shaping the meaning of risk. The structural dimensions of the risk society perspective are developed with reference to the operations of government, science and the legal system in defining the nature and the boundaries of risk. In Chapter 4, the pivotal role of the mass media in representing and communicating risk is discussed. At this stage, it will be demonstrated that the risk society thesis presents an impoverished account of the media, which underplays its centrality as a node of risk communication in contemporary society.

As an ensemble, the first four chapters of the book are broadly oriented towards understanding and evaluating the construction, production and mediation of risk. From Chapter 5 onwards, the axis of inquiry turns toward the way in which risks are comprehended, managed and consumed by individuals in everyday environments. In Chapter 5, we venture into the sticky area of risk perception, contrasting Beck's universal ideals with the heterogeneity of public understandings of risk. Applying the risk society thesis to existing empirical research, we argue in favour of a more fluid approach, which captures the culturally grounded fashion in which people negotiate risk. Chapter 6 examines the visible imprints made by risk, tracing the outcomes of structural shifts on the quality and diversity of everyday practices. In this chapter, priority is accorded to the material effects of risk and individualisation on the family, the workplace and personal relationships. In the penultimate chapter, the cognitive aspects of everyday risk negotiations are highlighted through consideration of the symbiotic relationship between trust, reflexivity and risk. Stepping beyond widely stated claims of public distrust in expert systems, we reconvene the evidence in order to promote a more conditional reading of expert–lay relations in contemporary society. Building on previous theoretical discussion of the constitution of reflexivity, in Chapter 8 we unload the relationship between risk consciousness, public debate and political transformation. In particular, current political trends will be assessed as a means of quantifying possible drift away from politics based on class, toward a 'politics of risk'. In order to investigate the potential of active micro-level disputes, patterns of political engagement will be related to the practice of 'subpolitics' (Beck, 1992: 183; 1999: 91). Using the political debate about genetically modified (GM) organisms as a touchstone, we interrogate the union established between risk and politics. At a theoretical level, the emancipatory trajectory of subpolitics is set against the restrictive capacities suggested by the governmentality approach. By adopting a strategy which incorporates both review and critique, I intend to construct an equitable appraisal of the risk society thesis; one which recognises its novel and progressive aspects, alongside the many theoretical holes and empirical oversights. If we are to develop a better understanding of how risk is represented, perceived and negotiated within everyday life, theory needs to be nudged ever closer to practice. Of course, this is an organic and processual activity which lies beyond as well as within the pages of this book.

1
Mapping the Risk Society

Since the mid 1980s, the concept of risk has acted as a fulcrum for the sociological project of Ulrich Beck. The seminal *Risk Society* (1992) has been widely acclaimed as the centrepiece of Beck's work. The book, which has sold well over 60,000 copies worldwide, propelled its author into the spotlight and produced significant reverberations, both within and outside academic circles (McGuigan, 1999; Rustin, 1994). *Risk Society* (1992) is amongst the most ambitious and provocative of texts written within the social sciences in recent years. Not only does it sweep through an extensive range of topics, it is also – by turns – furious, projective, ironic and humorous. Predictably, the book broke the mould within academic publishing, casting the writer forth as part heretic, part sociological clairvoyant. In his native Germany, Beck is esteemed not only for academic achievement, but also for his thought provoking contributions to high circulation newspapers and magazines (Bronner, 1995: 67). As McGuigan notes, the ripples produced by *Risk Society* (1992) have extended well beyond the confines of the university campus: 'It is not just a work of abstract social theory, but a significant intervention in the public sphere of the Federal Republic, a bestseller and required reading for the chattering class' (McGuigan, 1999: 125).

It should come as no surprise that the academic and social debate about risk has mushroomed since the publication of *Risk Society* (1992).[1] Reflecting on the author's background, one comes to understand the unprecedented breadth and diversity of the risk society approach. Beck's work is truly eclectic; he is 'the master of many traditions and the servant of none' (Bronner, 1995: 68). Through doctoral research, Beck developed an interest in the production of social knowledge and the application of science. In the risk society thesis, concern about the construction of scientific 'objectivity' is illuminated in the ecological problematique and consolidated through a sustained critique of expert systems. Beck's early research into industrial sociology and the sociology of the family finds voice in his analysis of structural changes in employment, family life and social relationships. In addition, German political conventions – in

11

particular, the politics of the Green movement – have sculpted the risk society perspective. As far as academic tradition is concerned, the concentration on social structure and the negative consequences of capitalist development are indelibly Germanic, following in the footsteps of Marx, Weber and Simmel. Within contemporary sociology, Beck's work has been contrasted with the drier and structurally inspired projects of Luhmann and Habermas (see Lash, 2002).

In a nutshell, Beck's groundbreaking approach charts the relationship between the unbinding of social structures, qualitative changes in the nature of risk and shifting patterns of cultural experience. Following on from the publication of *Risk Society* (1992), the concept of risk has remained an omnipresent feature of Beck's work (1994; 1995b; 1997; 1999; 2000b; 2002). In more recent offerings, Beck has continued to mobilise risk as an articulation point for debate about the restructuring of employment relations (2000a), the diversification of political activity (1997; 1999) the contents of globalisation (2000b) and the threat of international terrorism (2002). Despite focussing on an extensive range of subjects, Beck has reserved risk as a vital theoretical referent. In the midst of a peculiar mixture of acclaim and bitter criticism, Beck has consistently maintained that contemporary western society is embedded in a culture of risk which has profound impacts on the nature of everyday life.

In this opening chapter, I wish to provide an inventory of the theoretical materials used to assemble the risk society perspective. Firstly, the essential features of the argument will be recounted with reference to the 'pillars of risk' which prop up Beck's thesis. Secondly, the two rudimentary processes that propel the risk society – namely individualisation and risk distribution – will be outlined. By reviewing the defining features of the risk society, an appropriate theoretical framework for the issues discussed in subsequent chapters will be erected. Insofar as our review will remain exegetical, the pillars and processes of risk will be revisited with a more critical eye throughout the book. However, in order to appreciate the resonance of Beck's work, it is necessary to consider the broader concept of risk. Thus, prior to emptying out the risk society perspective, it may first prove instructive to briefly chronicle the semantic history of risk.

DEFINING RISK

As Lupton (1999a: 8) notes, 'risk' is a word that is commonly used to indicate threat and harm. In everyday parlance, the term 'risk' is used as 'a synonym for danger or peril, for some unhappy event which may happen to someone' (Ewald, 1991: 199). Whilst this definition is apposite to the modern age, it is important to recognise that the meaning of risk has evolved over time. In historical terms, risk is a relatively novel phenomenon, seeping into European languages in the last 400 years. However, there remains a distinct lack of consensus about the etymology of 'risk'. Some historians believe that the term derives from the Arabic word *risq*, which refers to the acquisition of wealth and good fortune (Skeat, 1910). Others have claimed that 'risk' finds its origins in the Latin word *risco* and was first used as a navigational term by sailors entering uncharted waters (Ewald, 1991: 199; 1993: 226, Giddens, 1999: 1, Lupton, 1999a: 5; Strydom, 2003: 75). Curiously, these two derivations of risk were soldered together in the seventeenth century through the principle of maritime insurance. Under this banner, risk came to relate to the balance between acquisitive opportunities and potential dangers (Wilkinson, 2001: 91). Such quantitative conceptions of risk were bolstered by the probability theorems of seventeenth-century mathematicians, such as Pascal and de Fermet (Bernstein, 1996: 68). The process of economic development in the eighteenth and nineteenth centuries further cemented risk to calculation. In this period, the steady expansion of industry and capital ensured that risk became associated with the economy through the activities of investors and bankers. By the beginning of the twentieth century, 'risk' was commonly used in insurance and finance to describe the possible outcomes of investment for borrowers and lenders. In modern times, risk remains firmly coupled to the economic world through forms of statistical calculation, stock-market speculation and company acquisitions. Capitalist markets cannot be sustained without risk, which is ingrained in the decisions of fund managers, the speculations of market makers, the borrowing of business managers and the valuations of insurance companies.

In economic and statistical terms, the concept of risk has traditionally been set apart from uncertainty. As Frank Knight explained many moons ago:

> The practical difference between the two categories, risk and uncertainty, is that in the former the distribution of the outcome in a group of instances is known ... while in the case of uncertainty this is not true, the reason being in general that it is impossible to

form a group of instances, because the situation dealt with is in a high degree unique. (Knight, 1921: 233)

The distinction between risk and uncertainty became received wisdom amongst insurers, economists, analysts and technical practitioners in the early and mid twentieth century. It would appear that many professionals – including those involved in insurance and health protection – still adhere to this logic in their decision making. Nevertheless, in contemporary society, the degree of overlap between risk and uncertainty militates against simple division. For example, the link between Bovine Spongiform Encephalopathy (BSE) in cattle and a new variant of Creutzfeldt-Jakob Disease (vCJD) in humans began as an uncertainty which transmuted into a risk. Seemingly unique cases of uncertainty can rapidly evolve into risks, as and when harm is established. Further, the BSE case illustrates that a high degree of uncertainty often surrounds the extent and the geography of harm (see Adam, 2000b: 119). Contemporary risks contain residual uncertainties, which render quantification problematic. At the height of the BSE crisis, expert estimations of future sufferers of vCJD ranged from 100 to 1 million (Hinchcliffe, 2000: 142). Given such a range of harm, distinguishing between risks and uncertainties seems a thankless – and a fruitless – task.

In modern society risk is inextricably linked to notions of probability and uncertainty. Theoretically speaking, a risk only arises when an activity or event contains some degree of uncertainty: 'the essence of risk is not that it *is* happening, but that it *might* be happening' (Adam and van Loon, 2000: 2). Today, the properties of probability and uncertainty are themselves welded to the idea of futurity. Risks are perceived to be hazards or dangers associated with future outcomes (Giddens, 1998: 27, Lupton, 1999a: 74). In modern discourse, risk relates to a desire to control and predict the future: 'To calculate a risk is to master time, to discipline the future. To provide for the future does not just mean living from day to day and arming oneself against ill fortune, but also mathematizing one's commitment' (Ewald, 1991: 207).

By unloading the composite features of probability, uncertainty and futurity we can begin to get a taste for the meaning of risk. However, at this stage, I do not intend to delve much deeper into the etymology of risk.[2] Indeed, 'defining' risk may prove to be something of a red herring. Firstly, understandings of risk will differ over time and place (Hinchcliffe, 2000; Lash and Wynne, 1992).

Secondly, the indeterminate character of risk ensures that perceptions will invariably be contested between individuals and social groups (Caplan, 2000; Fox, 1999). As Luhmann (1993: 71) points out, what are 'risks' for some, can be construed as 'opportunities' by others. Thirdly, in prioritising generality, catch-all definitions of risk tend to concede concrete meaning. Given the multidimensional nature of the concept, it is perhaps misguided to pursue a single definition. Putting aside epistemological issues, in closing down the meaning of risk we are liable to lose sight of the various takes on risk which illuminate Beck's project. When push comes to shove, compact definitions tell us little about the changing context of risk – or about how risk is constructed, interpreted and experienced through everyday interactions. Taking on board these qualifications, restricting our understanding of risk is contrary to the usage – and the spirit – of risk in the risk society thesis. Indeed, these provisos may go part way in explaining Beck's reluctance to provide a precise definition.[3] In many ways, the concept of risk is beyond concise articulation. Instead, we might profitably see risk as a container for a bundle of issues that are not readily disentangled:

> The concept refers to those practices and methods by which the future consequences of individual and institutional decisions are controlled in the present. In this respect, risks are a form of institutionalised reflexivity and they are fundamentally ambivalent. On the one hand, they give expression to the adventure principle; on the other, risks raise the question as to who will take responsibility for the consequences, and whether or not the measures and methods of precaution and of controlling manufactured uncertainty in the dimensions of space, time, money, knowledge/non-knowledge and so forth are appropriate. (Beck, 2000e: xii)

Acknowledging such networked complexities, we can recognise the polysemic quality of risk and approach the concept in an inclusive fashion. With this in mind, the first stage of our journey into the risk society perspective entails a review of the socio-historic context which risks both arise out of and shape.

EPOCHAL PHASES OF RISK

The risk society thesis makes two crucial propositions about the nature of risk in contemporary society. Firstly, Beck posits that the composition of risk has fundamentally mutated. Secondly, the increasingly hazardous quality of risk is said to have generated apocalyptic consequences for the planet. In both *Risk Society* (1992) and *Reflexive Modernization* (1994), the concept of risk is encased within the wider framework of an historical narrative. Broadly speaking, three distinctive epochs are recalled: 'pre-industrial society' (traditional society), 'industrial society' (first modernity) and 'risk society' (second modernity). The crux of Beck's argument is that changes in the composition of risk, allied to major structural trans-formations, have facilitated a transition from pre-industrial to industrial modernity and, latterly, into the risk society (Beck, 1995a: 78; Goldblatt, 1995: 159). To draw contrasts between different periods, paradigmatic forms of risk are described. Drawing upon a fairly light historical contextualisation,[4] Beck differentiates between 'natural hazards' and 'manufactured risks'. Natural hazards such as drought, famine and plague are associated with the pre-industrial period. At the level of risk consciousness, natural hazards were commonly attributed to external forces, such as gods, demons or nature (Beck, 1992: 98; 1995a: 78). In the period of industrial modernity – roughly encompassing the first two thirds of the twentieth century – natural hazards are steadily complemented by a growing set of humanly produced dangers, such as smoking, drinking and occupational injury. At this stage of development, a discrete pool of knowledge exists about how to regulate both natural disasters and man-made risks. This is evidenced by the applied practices of health and welfare systems, environmental agencies and insurance companies (Beck, 1992: 98). Finally, in the advance into the risk society, environmen-tal risks – such as air pollution, chemical warfare and biotechnology – prevail. These potentially catastrophic risks stem from industrial or techno-scientific activities and come to dominate social and cultural experience. For Beck, the risk society can be described as: 'A phase of development of modern society in which the social, political, ecological and individual risks created by the momentum of innovation increasingly allude the control and protective institutions of industrial society' (Beck, 1994: 27).

In stark contrast to natural hazards, manufactured risks are decision-contingent, endogenous entities which are generated by the practices of 'people, firms, state agencies and politicians' (Beck, 1992: 98). Because manufactured risks arise out of the developmen-tal processes of modernisation they can be seen as socially rather than naturally produced. Beck (1992: 21) ties resultant public cognition of risk to broader social transitions, such as globalisation, the individualisation of experience, the questioning of expert systems and the burden of identity construction. These all-embracing structural shifts are said to be steering western cultures toward a distinctive form of 'reflexive modernity' (Beck, 1994: 2). Circumventing the theoretical quicksand, reflexive modernisation refers to the way in which patterns of cultural experience are uprooted and disembedded by underlying changes in social class, gender, the family and employment. As the structural certainties previously provided by governing institutions evaporate, people are pressed into routinely making decisions about education, employment, relationships, identity and politics. Consequently, in reflexive modernity individuals assume greater responsibility for the consequences of their choices and actions. According to Beck, the changing nature of risk is intrinsically wedded to the broader process of reflexive modernisation. Far from remaining static, both the constitution and the effects of risk have fluctuated over time. In industrial society, via the process of socio-economic development, a coterie of unmanageable risks emerge. As scientific, technological and economic practices become entrenched as instruments of social progress, manufactured risks continue to appear as environmental 'side effects'. In the movement into the risk society – from the 1970s onwards in Britain and Germany – the unremitting production of environmental risks forces society to confront the harmful consequences of capitalist development (Beck, 1999: 9). Beck urges us to recognise that risks are no longer an inevitable and benign aspect of social development. Qualitative variations in the nature of risk mean that dangers produced by the system can no longer be contained. In contemporary western society, recurrent economic, scientific and technological expansion has resulted in the creation of 'serial risks' which breed with such intensity that existing mechanisms of risk management become swamped (Beck, 1996: 27).

In the risk society thesis a toolbox of concepts is employed to distinguish between manufactured risks and natural hazards. In particular, we can identify a trio of 'pillars of risk' which hold Beck's argument together, namely: transformations in the relationship between risk, time and space; the catastrophic nature of risk and the breakdown of mechanisms of insurance. In order to understand the foundations underpinning the risk society thesis and its working rationale, we relate each of these in turn.

THE RELATIONSHIP BETWEEN RISK, TIME AND SPACE

The changing dynamics of the relationship between risk, time and space is a central feature of the risk society thesis. Beck believes that pre-industrial and early industrial cultures were prey to forms of risk that were geographically and temporally contained (Beck, 1999: 143). Natural hazards such as earthquakes, famine and flooding are depicted as archetypal dangers faced in traditional cultures. Despite possessing substantial force and harbouring negative consequences, Beck contends that 'natural hazards' are temporally 'closed' phenomena which impact within a specific locale. Of course, natural hazards still threaten human life in many areas of the globe. However, the detrimental effects of natural hazards have been countered, managed and dissipated. In western cultures, natural hazards such as drought and famine have been all but banished (Giddens, 1991: 116). Further, the adverse effects of earthquakes have been lessened by construction restrictions in volatile areas and the production of shock-resistant buildings. From pre-industrial to industrial society, the incidence of economic and technological risks rises and accidents are recognised as the products of faulty human decisions (Beck, 1995a: 78). In the transition from pre-industrial to risk society, hazards and accidents become displaced by an aggregation of man-made risks. These socially produced risks are both more mobile and more oblique than preceding forms of danger. For Beck, the paradigmatic manifestation of environmental risk is the Chernobyl disaster (Beck, 1987; 1992: 7). In comparison with the natural hazards which typified pre-industrial life, the Chernobyl accident shattered geographical boundaries. The risks created by the reactor explosion were not tied to locale, endangering citizens far and wide. Toxins leaked from the plant not only caused ill health to citizens in Belarus and the Ukraine, but also glided over national boundaries, producing unknown effects (Beck, 1992: 7; Wynne, 1996: 62). Leaning heavily on the Chernobyl example, Beck postulates that environmental risks such as nuclear and chemical pollution remap the geography of risk. In the Chernobyl case, the safety procedure for nuclear accidents only covered a radius of 25 kilometres (Beck, 1995a: 78).

On top of transcending spatial limits, Beck asserts that the perils of the risk society cannot be temporally limited. The deleterious effects of environmental risks do not necessarily occur instantaneously (Adam and van Loon, 2000: 5). Years after the Chernobyl explosion, thousands of Ukranians and Belarussians developed serious

cancers and breathing disorders. Given that nuclear fallout is inde-
structible, toxins harboured underground and in the atmosphere
continue to produce damage to future generations: 'the injured of
Chernobyl are today, years after the catastrophe, not even all born
yet' (Beck, 1996: 31).[5] The sequential knock-on risks associated with
the Chernobyl accident bear testament to the seriality of manufactured
risks. As we shall see in Chapter 7, a decade after the reactor explosion,
radiocaesium related to chemical fallout from the Chernobyl accident
was discovered in Cumbria in the North-West of England (Wynne,
1996: 62). In the Cumbrian case, radioactive chemicals stored in the
ground generated a kaleidoscope of potential risks to the natural
environment, grazing animals and local residents.

Beck believes that incidents such as Chernobyl reformulate social
understandings of risk (Beck, 1995b: 504). The dangers affixed to
modern risks are not subject to temporal restrictions and defy
geographical enclosure (van Loon, 2000a: 169). Many threats are
invested with a capacity for destruction which is only partly
manifested on site (Beck, 1999: 3). To further shroud the situation,
prominent social risks – such as Aids, cancer or vCJD – cannot be
attributed to solitary sources (Smith, 2001: 157).

THE CATASTROPHIC NATURE OF RISK

Besides possessing temporal and spatial mobility, manufactured risks
are said to be more catastrophic than the natural hazards which
prevailed in previous eras (Beck, 1995a: 100). To illustrate the
catastrophic nature of risk, Beck habitually refers to three 'icons of
destruction': nuclear power, environmental despoliation and genetic
technology (1992: 39; 1995a: 4). It is argued that each of these technological forces has the potential to yield a 'worst imaginable accident'
(WIA), resulting in the extinction of human life (Beck, 1999: 53). In
recent times, biotechnologists have uncovered the scientific formula
for reproductive cloning, with rogue scientists grotesquely competing
to create the first genetic replicate (Follain, 2001: 30). In the case of
nuclear power, a single atomic explosion may be sufficient to
annihilate civilisation, shattering established barriers of risk
management, insurance and aftercare (Irwin et al, 2000)
Unquestionably, the safety of nuclear technology is testable only
after it is manufactured. Such a definitively high-risk climate leads
Beck to conclude that technological modernisation has transported
society to the brink of self-destruction:

There would probably soon be unanimous agreement on one basic historical fact: namely, that the second half of the twentieth century has distinguished itself – by virtue of the interplay of progress with the possibility of annihilation by the ecological, nuclear, chemical and genetic hazards we impose upon ourselves. (Beck, 1995a: 83)

Beck avers that it is now possible for humans to destroy *all* that we have created, with *what* we have created (Lupton, 1999b: 4). Not only are contemporary environmental threats potentially devastating, they also produce cumulative effects (Caplan, 2000: 167). The massification of risk is aptly illustrated by the threat to public health currently posed by poisonous substances in the atmosphere. Arguing along paths well trodden by ecologists, Beck contends that the toxicity of contemporary forms of environmental pollution is much greater than in either pre-industrial or industrial modernity. As we shall see in Chapter 2, capitalist mass production and technological development have resulted in an irreversible extension of toxins in the atmosphere. Relying heavily on the three icons of destruction, Beck contends that western cultures are currently living under the penumbra of self-annihilation via the manifestation of worst imaginable accidents.

THE BREAKDOWN OF SOCIAL INSURANCE

The third pillar of risk relates to the ability of social institutions to ensure public safety. Throughout his work, Beck (1992; 1995a; 1998) implies that the unmanageable quality of manufactured risk has adversely impacted on the social institutions charged with maintaining health and security. Of particular interest is the manner in which governing institutions have traditionally deciphered liability claims and organised compensation packages. This historical aspect of the argument is most fully developed in *Ecological Politics in an Age of Risk* (1995a). In this text, risk assessment and liability procedures in industrial modernity are pitted against those applied in the present day. Elaborating on the framework developed in *Risk Society* (1992), Beck contends that the rate of technological development is growing exponentially, leading to the heaping of environmental risk. The scope and prevalence of damaging side effects presents problems for public institutions responsible for insuring against risk. Again, the explosive hazards of the risk society are contrasted with the more benign and institutionally manageable risks of the simple industrial epoch. Notably, Beck has rather little to say about methods of risk calculation in pre-modern society. This omission is indicative of a broader tendency to shift between analyses of different epochs without adequately explaining the bases of distinction.[6] Nevertheless, reading between the lines, Beck implies that organised systems of insurance were absent in pre-modern cultures and that methods of conceptualising hazards incorporated fate and blame. Pre-Enlightenment, Christian understandings of existence typically featured a hierarchical and immutable 'chain of being'. Assuming that social consequences were largely understood in relation to fate, it is probable that individuals will have perceived danger as an unpleasant but unavoidable aspect of life.

Post-Enlightenment, with the development of a secular evolutionist cosmology, citizens increasingly demanded both explanations and compensation for risks. These demands were institutionalised in the nineteenth century via the development of the 'safety state' (Beck, 1995a: 107; Ewald, 1986). Beck contends that the gradual development of welfare systems within nation states was based upon two common goals. First, welfare systems acted as an antidote to the inevitable problems produced by rapid technological, economic and social change. Second, the formal welfare state provided citizens with a vehicle for processing various safety and security needs. In response to citizenship demands, organised systems were developed in health and welfare, the economy, law and insurance. The legitimacy of government thus became dependent upon the ability of the state to fulfil security pledges (1995a: 109). Of course, such an unqualified narrative of the history of the western welfare system is disputable. For example, far from representing a commitment to security, other theorists have maintained that western welfare systems were introduced to enhance social order and to defuse class tensions (Goldblatt, 1995: 168; Hillyard and Percy-Smith, 1988; Offe, 1984).

Rather than engaging with competing explanations, Beck instead maps out the welfare functions of structures of risk management. Playing off Marxist terminology, the institutions involved in the assessment and management of risk are referred to as the 'relations of definition' (Beck, 1995a: 116). The relations of definition are made up by an arsenal of institutions – such as government, the civil service, the legal system and scientific organisations – which produce 'the rules and capacities that structure the identification and assessment of environmental problems and risks' (Goldblatt, 1995: 166). As will be explained in subsequent chapters, the multiple functions of the relations of definition are absolutely crucial in informing and moulding public understandings of risk (Adam and van Loon, 2000: 4). In industrial society, the relations of definition are seen to be capable of managing issues of risk liability and risk compensation. During this period, insurance claims were resolved in relation to an actuarial system. By utilising calculative methods of assessment, the relations of definition acquired expertise and legitimacy in matters of risk. The probability of a risk materialising could be statistically calculated, appropriate compensation measures employed and, if necessary, the guilty party could be penalised through legal sanctions. Therefore, in the early and mid parts of the twentieth century, the risks and dangers produced by modernisation could be adequately managed by existent systems of causality, liability and insurance. To all intents and purposes, the quality of hazards produced in simple industrial society complemented the methods of risk assessment and management: 'if a fire breaks out, the fire brigade comes; if a traffic accident occurs, the insurance pays' (Beck, 1995a: 85).

This relatively secure system of risk management begins to transmute as society moves into the transitional period between industrial society and the risk society. During this phase, western cultures become beleaguered by intrinsic risks that welfare systems cannot properly address or eliminate.[7] On the cusp of the risk society, the natural hazards common to earlier epochs are complemented by a feral collection of risks which transcend traditional boundaries of time and space. In the break into the risk society proper, manufactured risks swell and multiply, revoking existing principles of liability:

> In all the brilliance of their perfection, nuclear power plants have suspended the principle of insurance not only in the economic, but also in the medical, psychological and cultural sense. The residual risk society has become an uninsured society, with protection diminishing as the danger grows. (Beck, 1992: 101)

Sidestepping several delicate phenomenological issues, the separation imposed between manufactured risks and natural hazards is unequivocal.[8] In contemporary culture, the continued development of non-limitable catastrophic risks implies that social institutions are unable to manage risk. Beck postulates that existing systems of civil liability are designed to deal with accidents and injuries of undisputed origin, such as acts of violence. Such dangers generally involve identifiable injuries, victims and guilty parties. Given the oblique and global nature of manufactured risks, such forms of legislation are no longer functional or appropriate.

To summarise, Beck contends that the nature of risk has altered dramatically over time. In pre-modern societies, natural hazards such as disease, drought and famine were prevalent. In contemporary western society, many of these basic risks have been overcome, being superseded by an unceasing collection of manufactured risks. At first glance, WIAs might be mistaken for remote and distant risks. However, the tentacles of manufactured risk protrude into countless cultural spheres, such as food consumption, leisure, sexuality and employment. As western cultures enter into the risk society, the institutional mechanisms for handling risks falter, producing systemic crisis. Effectively, manufactured risks overrun methods of insurance and compensation, engendering a crisis of institutional accountability (Beck, 2000d: 224). As we shall observe in Chapter 3, this enforced institutional intransigence leaves the relations of definition with little option but to engage in ineffective dramaturgical displays of risk management. The somewhat farcical result of institutional 'intervention' is that – instead of being curtailed – environmental risks simply proliferate. As a result, in the risk society, uncertainties are amplified and risk-regulating institutions become vulnerable to public scepticism and doubt (Caygill, 2000: 167).

RISK DISTRIBUTION AND TRANSITIONAL LOGICS

Having described the three pillars of risk on which the risk society thesis is built, in the remainder of the chapter I wish to provide a capsule account of individualisation and risk distribution. These two processes are fundamental to Beck's argument, being widely identified as the key drivers of the risk society (Engel and Strasser, 1998; Goldblatt, 1995; Mol and Spaargaren, 1993). In effect, the risk society perspective houses two interlinked theses: one relating to risk, the other pertaining to individualisation (see Lash, 2002). In Germany, the individualisation thesis has attracted attention and directed social debate. Meanwhile, in Britain, America and Canada the risk thesis has captured the academic imagination. Considering the centrality of each process to the risk society thesis, we will return to cast a more discerning eye over individualisation and risk distribution. For now, I simply wish to sketch out the essential characteristics of each process as a basis for specific application in forthcoming chapters.

In *Risk Society* (1992), Beck surmises that the creeping transition from industrial modernity towards the risk society has profound implications for the 'distributional logic'. This concept is rather nebulously defined by Beck, who employs the term to encapsulate a series of socio-political changes. The distributional logic can be broadly conceived as an overarching set of organising principles and practices which govern patterns of social welfare and the allocation of resources. These principles have ideological connotations and are coupled up to public perceptions of politics. The distributional logic also has a material dimension, involving the dissemination of the 'cake' produced by economic and technological development (Beck, 1992: 19). Hence, the distributional logic refers to a material process of distribution shaped by politics and social policy. Beck (1992: 23) believes that the distributional logic governs the provision of positive 'goods' such as health, wealth and educational opportunities and negative 'bads' such as risk, infection and disease. In industrial societies, political debate in the West revolved around the efficacy of the distributional logic in meeting the diverse needs of individuals and social groups. In short, the system's ability to trump up with and share out the goods. However, in the risk society, manufactured risks grow and spread, destabilising social goals and political priorities.

Beck (1992: 36) fixes underlying shifts in the distributional logic to particular epochs. The distributional logic of simple industrial society is explicitly bound to the distribution of social goods (Beck, 1992: 20). The developmental dynamics of industrial societies are fundamentally indexed to the ideal of equality, with the prime social objective being the satisfaction of basic material needs and a rising standard of living. In this sense, industrial society can legitimately be described as a 'class society' (Beck, 1992: 34). In industrial society, dominant concerns revolve around the distribution of wealth, ownership of material goods, equal opportunities and job security (Beck, 1992: 49).

The opposition between the industrial (class) society and the risk society is based around a binary distinction between scarcity and insecurity (Beck, 1992: 20). Whilst industrial societies are characterised by scarcity, the risk society is distinguished by insecurity (Scott: 2000: 34). In class societies unjust distribution is apparent and socially observable: 'Misery needs no self-confirmation. It exists ... the certainties of class societies are in this sense the certainties of a culture of *visibility*: emaciated hunger contrasts with plump satiety; palaces with hovels, splendour with rags' (Beck, 1992: 44).

In industrial society, malfunctions in the distribution of social goods – be they deliberate or accidental – cannot be kept secret. As the frequency with which manufactured risks appear quickens, social and political concerns alter and the distributional logic transmutes. However, in the movement from industrial to risk society, the distributional logic no longer revolves around how the 'cake' might be divided up. Instead, it becomes clear that the cake has become poisoned (Beck, 1992: 19). Through the production of risky products and unpalatable side effects, the cornerstones of modernisation – mass production, consumerism and techno-scientific development – are undermined and dislodged. In a hostile political climate, conflicts emerge around issues of hazard attribution, technical liability and institutional management: 'They erupt over how the risks accompanying goods production – nuclear and chemical mega technology, genetic research, the threat to the environment, over-militarisation and the increasing immiseration outside of western industrial society – can be distributed, prevented, controlled and legitimised' (Beck, 1994: 6).

By definition, global environmental problems cannot be properly regulated at a national level. In as much as industrial goods-distributing societies were bounded entities, the generation of social bads undoes the regulatory power of the nation state. Contemporary risks cross international borders, disembedding and unsettling political and economic interests (Lash and Urry, 1994: 33).

Historically, the issue of wealth distribution has – due to sustained pressure from the labour movement – been formally inscribed into the manifestos of western democratic parties. In recent times, successful political parties have come to power on a promise of enhanced wealth distribution through improvements in health, education and employment. Whilst such promises do not appear to have significantly closed the class divide, liberal democratic parties have paid lip service to the theoretical logic of more equal forms of wealth distribution.[9] However, Beck believes that undivided emphasis on the distribution of social goods is politically misguided, pronouncing that political energies should be redirected towards the elimination of social bads. In the first instance, western societies already cater for the basic survival needs of the vast majority. Moreover, even allowing for the egalitarian distribution of resources, contemporary western society would still fail to meet the physical and existential security needs of its citizens. On these grounds, Beck questions the political dominance of the logic of wealth distribution, calling for a new 'politics of risk' (Beck, 1996a; 1997; 1999). It is argued that the escalation of risk within modern culture has fuelled a discernibly negative and defensive political logic. As will be scoped out in Chapter 8, the production of extraordinary and catalytic risks forces political parties into unknown discursive territory. Politicians schooled in the principles of goods distribution are rendered inert by indeterminate and incalculable bads. Underneath a veil of intransigence and denial, politicians continue to pursue the agenda of social goods, imprudently tiptoeing around the burgeoning problem of bads. On the horizon, the generation of unfamiliar risks – dirty bombs, ebola, nanotechnology – continues unabated, redefining cultural values and political expectations. As we shall see in Chapter 6, this shift from class relations to risk relations generates new antagonisms between those who produce and those who consume risk (Beck, 1997: 19).

In many respects, Beck's theory of distributional logics marks a significant departure within social theory (Lash and Wynne, 1992). The macro theories of the 'founding fathers' of sociology were wrapped around the issue of goods distribution. Both Marx and Weber sought to reveal how, in capitalist cultures, socially produced wealth could be distributed in an unequal fashion. In contrast, rather than deploying class or status as the key determinant of social experience, the risk society thesis posits that a universalising process of distribution has loosened the ties of class-based identities and collective social experience. In order to illuminate different patterns of risk distribution Beck again reverts to the three-stage transitional model. In the industrial society, the production of social goods dominates social and political life. In the transitional period between the industrial society and the risk society, concerns about social goods are augmented by social bads, with the latter in the ascendant (Beck, 1992: 20). Finally, as society is catapulted into the risk society, the material effects of manufactured risk become unavoidable and social bads dominate the political landscape: 'the social circumstance which matters most in our intolerably jumbled modern condition is risk: all of us who inhabit the earth at the end of the twentieth century – rich and poor, high and low, young and old – live equally in the embrace of the risk society' (Jasanoff, 1999: 136).

For Beck, the current clash between the positive logic employed in wealth production and the negative logic of risk breeds confusion, distrust and insecurity. Living with risk means a radical change in normative values and social expectations. To substantiate this line of reasoning, the experiential aspects of 'class positions' are contrasted with 'risk positions' (Beck, 1992: 53). Class positions relate to the ability to attain socially produced wealth, whereas risk positions refer to the probability of exposure to risk due to social, economic and geographical circumstance. As manufactured risks intensify and spread, class positions are outmoded by risk positions, with the distributional axis revolving around safety, not equality (Beck, 1992: 38). In effect, social bads reformulate the nature of social and political conflict, generating 'safety needs' for minimisation, prevention and monitoring of risks. Because class logics are defined primarily by ownership, they invariably produce social unrest. Hence, in theory at least, class positions are open to question and reformulation via conflict between the resourced and the resourceless. However, relatively direct forms of argumentation are blown away by global environmental risks which uproot social hierarchies and reorganise the configuration of conflict: 'anyone affected by them is badly off, but deprives the others, the non-affected, of nothing' (Beck, 1992: 40). To varying degrees, citizens are enabled by the distribution of goods to possess differing degrees of wealth. By contrast, nobody seeks ownership of social bads (Beck, 1992: 23). Hence, the logic of the risk society is based not on possession, but avoidance:

> The dream of class society is that everyone wants and ought to have a *share* of the pie. The utopia of the risk society is that everyone should be *spared* from poisoning ... the driving force in the class society can be summarised in the phrase *'I am hungry!'* The driving force in the risk society can be summarised in the phrase *'I am afraid!'* (Beck, 1992: 49)

Whilst in industrial society wealth and risk distribution coincide, in the risk society the logics are prone to bifurcation. The changing nature of environmental dangers effectively skews the neat fit between class and risk. Because modern businesses and institutions act as conduits for global risks, even the affluent are threatened. To explain this phenomenon, Beck (1992: 37; 2002: 4) discusses the impacts of 'boomerang effects' which transcend traditional boundaries of class and nation, catching up with those who generate or produce profit from them. As we shall see in Chapter 2, environmental pollution is a prime example of the universalising boomerang effect, threatening capital accumulation, throwing expert systems into turmoil and posing global health risks.

INDIVIDUALISATION: THE UNBINDING OF EXPERIENCE

Having sketched out discernible changes in the distributional logic, I wish to conclude the chapter by briefly prising open the relationship between individualisation and social structure. In the risk society, common associations, social structures and patterns of identity are cast into flux and subject to doubt. According to Beck, these far-reaching changes are catalysed through the individualisation of experience. Condensed down, individualisation refers to the way in which everyday interactions have become de-traditionalised due to shifts in social structure and patterns of enculturation: 'Biographies are removed from the traditional precepts and certainties, from external control and general moral laws, becoming open and dependent on decision-making, and are assigned as a task for each individual' (Beck and Beck-Gernsheim, 1995: 5).

Beck believes that contemporary self-identities are constructed, mobile entities which cannot be dissociated from individual choices. As will be illuminated in Chapter 6, traditional gender roles, the nuclear family, sexuality and employment paths all become open to question, decision and modification (Beck and Beck-Gernsheim 1995: 6). In this respect, Beck's work fits hand in glove with that of the British sociologist, Anthony Giddens (1990; 1991; 1994).[10]

Beck tells us that in contemporary western society everyday experience of risk has become individualised. The gradual loosening of the structures and networks of tradition forces social actors to confront risks as individuals rather than as members of a collective (Furlong and Cartmel, 1997: 27). The effects of individualisation have been accentuated by globalisation, the erosion of the welfare state and the failure of the relations of definition to regulate risk (Beck, 2000: 21). The order and predictability which prevailed in industrial societies give way to the unstable and chaotic experience of the risk society. Although we might reasonably expect each generation to experience a different cultural reality, Beck (1998: 11; 1995a: 17) believes that the frenetic lifestyles of young people – whom he poetically refers to as 'freedom's children' – provide the clues that a historical watershed is taking place.

In *Risk Society* (1992), it is argued that the flexibilisation of the labour market, the decline in manufacturing and heavy industry, educational differentiation and changing patterns of consumption have dissolved the bonds of collective experience, leading to atomised forms of existence. Where industrial society is structured through the composition of social classes, the risk society is individualised (Beck, 1992: 91). Beck believes that a social logic which focusses solely upon the issue of wealth cannot properly address the fragmented concerns and dilemmas currently faced by social actors. In effect, the logic of wealth distribution is misaligned with the habituated practices of risk management, decision and choice undertaken by individuals in western society.

CONCLUSION

To recap, Beck hypothesises that the social cleavages which inhere in the risk society cannot be comprehended using the traditional sociological categories of class, gender and age. The emergence of boomerang effects signifies that previously defined boundaries between those who gain and those who lose from risk have been blurred. Given the lingering spectre of WIAs, everyone ends up residing in the latter camp. In the early stages of transition from industrial to risk society, risk and class positions tend to merge, leaving the materially disenfranchised most endangered by risk. As society enters the risk society proper, class-versus-class cleavages are steadily replaced by risky conflicts between various groups and sectors. As the production of manufactured risks increases, the distributional logic within society mutates. This calendrically indistinct point signals that risk positions no longer relate exclusively to wealth, but are also mediated by fate and geographical location (Beck, 1995a: 154). In Beck's estimation, the individualisation process and changing patterns of risk distribution herald a radical restructuring of political ideology. At an everyday level, the processes of individualisation and risk distribution force social actors to confront a plethora of uncertainties, hazards and choices.

Naturally, our attention in Chapter 1 has been inflected toward the pillars of risk which support Beck's argument and the macrosocial processes which catalyse the risk society. Through repeated reference to the icons of destruction the risk society thesis portrays a vision of society in which the darkening cloud of nuclear, chemical and genetic technologies endangers human existence. For Beck, the appearance of catastrophic environmental dangers implies that the rationalisation of risk as an exogenous entity is no longer a viable option. The experience of contemporary risk is not simply about fear at a distance. Rather, a fleet of risks have seeped through into everyday experiences of work, friendship and the lived environment.

2
Risk and the Environment

Hitherto we have laid out the components of the risk society thesis, homing in on the defining processes of individualisation and risk distribution, the pillars of risk and icons of destruction. Setting aside the processes of individualisation and risk distribution for the moment,[1] in Chapter 2 I wish to begin to test the reliability of the three pillars of risk on which the risk society thesis is built, using the environment as a focal point for debate. We begin by presenting an overview of Beck's case, establishing the ways and means by which environmental threats are routinely produced in contemporary capitalist societies. This discussion paves the way for a more detailed account of the inability of institutions to diminish hazard production and the subsequent rise in public anxiety about environmental risks. In later sections, we formally commence our critique of the risk society perspective, questioning the crude differentiation between risks and hazards, the decidedly patchy use of empirical evidence and the sanguine portrayal of the catalytic political power of environmental threats.

In Beck's design, the environment is cast as a terrain which is indelibly inscribed by systems-generated risks. Both *Risk Society* (1992) and *Ecological Politics in an Age of Risk* (1995a) contain animated, enraged tracts on the destructive effects of capitalism on the natural habitat. The former describes how industrialisation and modernisation have generated uncontrollable threats to the environment. In the latter, Beck explains how the incapability of western institutions to regulate techno-scientific development has led to a proliferation of ecological dangers. Without doubt, environmental risk is an indispensable ingredient in the risk society mix:

> Beck's sociology and the societies it describes are dominated by the existence of environmental threats and the ways we understand and respond to them. Indeed, one could go as far as to argue that the risk society is predicated on and defined by the emergence of these distinctively new and distinctively problematic hazards. (Goldblatt, 1995: 155)

30

In his homeland, Beck's ideas have been championed by the Green movement. A longstanding environmental campaigner, Beck is a former member of the influential Future Commission of the German government and is rumoured to be an informal advisor to Chancellor Gerhard Schröder. In the narrower corridors of the academy, the risk society thesis has acted as a source of stimulation and inspiration for those concerned about ecological affairs (see Goldblatt, 1995; Marshall, 1999; Strydom, 2003). Despite receiving plaudits for his eco-political trajectory, Beck's understanding of the relationship between nature and culture has attracted a decidedly glacial response (see Dickens, 1996; McMylor, 1996; Smith et al., 1997). Reflecting these disparate traditions, here I wish to afford due weight to the progressive aspects of the risk society perspective, as well as fetching up prominent areas of theoretical weakness.

THE PRODUCTION OF ENVIRONMENTAL RISK

The risk society thesis expresses major apprehensions about the effects of economic and techno-scientific development on the lived environment:

> Our epoch has taken progress so far that a minimal exertion may relieve everyone of all further exertions ... we have done away with life after death, and placed life itself under permanent threat of extinction. Nothing could be more transient. Yet we have done more: we have elevated transience to a principle of progress, released the potential for self destruction from its restriction to warfare, and turned it, in manifold terms, into the norm: failsafe and ever more failsafe atomic power plants; creeping and galloping pollution; the latest creations of genetic engineering, and so forth. (Beck, 1995a: 4)

In addition to the catastrophic potential of worst imaginable accidents, Beck also expresses deep concern about the snowballing impacts of capitalist development on the environment. In *Risk Society* (1992), the damaging ecological effects of economic and scientific development take centre stage. On occasion, Beck outlines the negative consequences of ongoing environmental processes, such as global warming and air contamination. In other passages, particular environmental disasters, such as Chernobyl and Bhopal, are employed as vignettes of risk. At the heart of the risk society approach lies a conviction that the increasing frequency with which environmental risks are produced – coupled with their palpable effects on the environment – signifies a movement from a relatively safe and ordered industrial society, to an insecure and fragmented risk society.

Beck believes that the historically assumed boundary between the natural and social has been eroded by relentless technological advances, scientific applications and economic development. In previous epochs, the realms of culture and nature were discrete. In modern society, nature has become thoroughly humanised: 'the risk society begins where nature ends' (Beck, 1998a: 10). It is argued that, in capitalist societies, customary activities and practices have produced a boundless collection of environmental 'bads', such as air pollution, global warming and acid rain. To prop up his analysis, Beck earmarks two distinctive trends arising out of the ever-closing relationship between humans and the environment. Firstly, at a practical level, the pervasive geographical span of environmental dangers demonstrates that locally produced risks create global consequences. Secondly, at a perceptual level, the escalating scale of environmental problems leads to greater social awareness of the harmful impacts of human practices on the flora and fauna of the planet. These two features indicate that risk is becoming a universal issue, both in terms of environmental impact and social cognition. In stark contrast to the predictable layering of poverty, the distribution of environmental risk is open to chance. In this sense, environmental risks are theoretically egalitarian. Assuming that we all breathe, eat and drink, everyone and anyone can be exposed to environmental risk: 'reduced to a formula: poverty is hierarchic, smog is democratic' (Beck, 1992: 36). Beck believes that risks are no longer perceived as fated and unavoidable accidents visited upon society by nature. Instead, in contemporary culture, risks are related to human actions and (non) decisions (Beck, 1995a: 2).

In the risk society, the unrelenting generation of environmental dangers undermines scientific and political authority and untacks social structure. Chemical, nuclear and genetic threats rearrange conventional hierarchies of class, gender, and geography (Beck, 1997: 159). Risk distribution in industrial society chases hierarchical class structures, whereas modern risks follow a circular motion which transcends established patterns of affluence and poverty. In contemporary risk society, the categories of perpetrator and victim dissolve (Beck, 1992: 38). Because 'mega-hazards' stretch the boundaries of time and space, risks have the capacity to revisit wealthy risk-generating nations. In the final analysis, insuring against exposure to environmental risks is impossible, even for those with status and economic power: 'there are no bystanders anymore' (Beck, 1996a: 32).

RISK REGULATION: A CASE OF INSTITUTIONAL PARALYSIS?

The risk society narrative of environmental danger describes a disparity between the nature of manufactured risk and the institutional apparatus responsible for risk regulation. It is argued that the existing relations of definition were set up to deal with relatively undisruptive and predictable risks. Although social institutions functioned effectively in the early stages of capitalist development, regulatory agencies have become stymied by unpredictable environmental hazards. In effect, nineteenth-century methods of security are being applied to twenty-first-century risks. Again, Beck draws upon a series of illustrations and examples to corroborate his claims (Beck, 1995a: 59; 1998a; 1999: 155). The global appearance of manufactured risks throws customary procedures of risk calculation and accountability into sharp relief. In contradistinction to systems of attribution in pre-modern and industrial cultures, nobody appears to be individually responsible for environmental risks. In many instances, questions of liability and compensation are obscured by the indeterminate and multicausal nature of risk production (Macnaghtan and Urry, 1998: 106; Smith, 2001). As a result, searching for culpable parties is akin to identifying the toxic vegetable in a *pot-au-feu*: 'No one any longer has privileged access to the uniquely correct way of calculation, for risks are pregnant with interests, and accordingly the ways of calculating them multiply like rabbits' (Beck, 1995a: 135).

The incapacity of the extant relations of definition to regulate against risk is highlighted by the continued production of environmental pollution. At well publicised global summits, scientific experts and leading politicians discuss 'controllable emissions' and decide upon 'acceptable levels' of air pollution. The existing process of risk regulation very much depends upon the success of international legislation in preventing threats to public health arising from poor air quality. Should acceptable regulatory levels be broken, guilty parties can be brought to court and punished by legal sanctions. Furthermore, victims of environmental risk may be awarded compensation and sentences passed act as a warning to other would-be offenders.

However logical in theory, the practical application of the regulatory process is obstructed by the widespread and generalised production of environmental risks. Beck contends that power holders are determining levels of pollution using methods of risk assessment which have long since become defunct. Given that global risks have immeasurable and indeterminate effects on quality of life, statistical estimates of 'acceptable levels' of pollution are meaningless, 'at least as long as "safety" or "danger" has anything to do with the people who swallow or breath the stuff' (Beck, 1992: 26). In the case of air contamination, the very idea of 'acceptable levels' puts environmental despoliation up as an acceptable trade for profit maximisation. The ideological connotations of expert discourses of regulation are not lost on Beck:

> The subject of this decree then, is not the prevention of, but the *permissible extent* of poisoning. *That* it is permissible is no longer an issue on the basis of this decree ... the really rather obvious demand for non-poisoning is rejected as *utopian.* At the same time, the bit of poisoning set down becomes *normality.* It disappears behind the acceptable values. Acceptable levels make possible a *permanent ration of collective standardized poisoning.* (Beck, 1992: 65)

By persisting with outmoded quantitative methods of risk analysis, scientific and technical experts have effectively reached a plateau of paralysis. The social outcome of this is that pressing public questions about safety, regulation and responsibility for environmental risk remain unanswered by state institutions. Concomitantly, the incessant appearance of manufactured risks illuminates the inefficacy of the traditional logic of identifying isolated sources. Environmental risks are invariably multi as opposed to monocausal:

> It is obviously impossible to bring individual substances into a direct, causal connection with definite illnesses, which may also be caused or advanced by other factors as well. This is equivalent to the attempt to calculate the mathematical potential of a computer using just five fingers. Anyone who insists on strict causality denies the reality of connections that exist nonetheless. (Beck, 1992: 63)

By persevering with archaic methods of risk assessment, the relations of definition are inadvertently promoting the ongoing manufacture of threats. Following prescribed rules of risk regulation, ill effects need to be proven by affected parties. In short, the responsibility for establishing harm rests with the victim. The legal requirement to establish a causal relationship between a specific ailment and the production of pollution effectively means that risk generators are privileged in law. Moreover, given that prevention of environmental risk would require proof of pre hoc rather than post hoc toxicity, legal procedures are powerless to avert the future emergence of environmental hazards. Under current legislation, a direct causal relationship between ill health (such as asthma, bronchitis, lead poisoning, cancer) and particular environmental risks (such as contaminated air) is notoriously difficult to prove in a court of law. Firstly, pollution may actually have been produced miles away from the site of contamination. Secondly, once in court, the legal requirement to establish individual culpability enables defendants to blame other parties, particularly in heavily industrialised areas (Beck, 1995a: 135). To further unbalance the scales of justice, individual victims seeking to prove toxicity invariably have fewer resources, reduced access to vital information and less knowledge about the workings of the legal system than large profit-rich companies which invariably constitute offending parties. In this way, 'proof' evades the grasp of victims of environmental risk and air quality continues to deteriorate (Goldblatt, 1995: 155). In the end, the legal stakes are institutionally loaded in favour of the polluter and culpability for environmental risk is anonymised. The absurdity of the situation is not lost on Beck: 'note the consequence: the pollutants pumped out by everyone are pumped out by no one. The *greater* the pollution, the *less* the pollution' (Beck, 1995a: 135).

Beck avers that the doubt and uncertainty surrounding environmental risk production have enabled guilty parties to eschew the burden of blame. In 1985 the German legal system investigated 13,000 incidents of environmental contamination. Of this number, only 27 convictions with prison terms were secured, 24 of which were suspended and three of which were dropped (Beck, 1995a: 134). Although such statistics might be read off as indicators of a lenient criminal justice system, Beck believes that failure to establish responsibility for environmental risk is systematic and bears testament to the dysfunctional structure of the legal system. Since legal liability is predicated on the tenet of absolute proof, 'holding a single individual liable is comparable to trying to drain the ocean with a sieve' (Beck, 1995a: 2). Technically speaking, institutional incapacity renders the risk society a habitat beyond insurance and liability (Adam and van Loon, 2000: 7).

In the risk society narrative, the inability of the relations of definition to curb risk is depicted as systemic, rather than erratic. However, in light of a lack of viable alternatives, tried, tested and failed modes of risk management continue to be routinely trotted out by safety-critical institutions charged with maintaining public safety. In the short term, a concatenation of guilty parties escape liability and punishment for the production of environmental dangers. In the long term, manufactured risks are rendered invisible, exacerbating ecological problems.

ECOLOGICAL POLITICS IN THE RISK SOCIETY

Although we will tackle the politics of risk head on in Chapter 8, it is worth briefly staking out Beck's claims about the political consequences of environmental risk production. In this section we pull out some of the alleged consequences of environmental risk generation for levels of public engagement. Beck believes that the autocratic and Byzantine nature of institutional risk regulation has restricted public involvement in the decision-making process. Of course, the absence of public involvement in social decision-making is an issue which has been widely documented elsewhere (see Ho, 1997; Woollacott, 1998). However, Beck's work is distinctive in that it turns environmental risk into a conductor for political engagement. The risk society perspective assumes that public interest in environmental issues rises as manufactured risks leak into the public sphere – both in the literal and the informational sense. The visibility and frequency of environmental problems leads to increasingly inquisitive public attitudes toward political power holders and decision makers. Although environmental risks are not directly borne by the political process itself, citizens expect the state to take action to protect public health (Giddens, 1998: 29). However, manufactured risks such as genetic cloning and bioterrorism transgress territorial boundaries and escape the regulatory powers of national governments, meaning that crucial political and environmental issues can no longer be solved within isolated nation states. Although the risk society critique is primarily levelled at the extravagances of western capitalism, Beck is equally alert to the dismal ecological record of communist regimes. Both free-market capitalism and state-centred communism stand accused of failing to halt the propagation of environmental risks. In the light of the political failures of left and right, the risk society thesis calls for systematic political restructuring to realign institutional structures with contemporary global trends: 'The nation state has lost any capacity to act on the problems that are moving the world, from environmental protection to global economic interconnections and migration to issues of regional and global peacekeeping' (Beck, 1998a: 107).

In *World Risk Society* (1999), Beck differentiates between 'simple globalisation' or top-down politics and 'reflexive globalisation' which works from the bottom up. The former model merely extends the existing power base, with transnational power blocks such as G8, the World Trade Organisation and NATO doing little to limit the diffusion of environmental dangers. In contrast, Beck believes that reflexive globalisation has the potential to fundamentally challenge the institutions and agents which generate environmental risk. Through engaging in ecologically sound practices and applying political pressure, the public have begun to reshape society from the bottom up, leading to the ad hoc democratisation of social criticism and political decision making (Beck, 1998a: 37). Various forms of 'subpolitics' have emerged in conflictual spaces, where members of the public have used oppositional political values as a basis for active debate and protest. As globalisation forces the locus of political decision making to mutate, local actions produce global impacts; petition signing, local campaigning, protest marching and boycotting products all act as contemporary methods of 'direct balloting' (Beck, 1999: 42). Accordingly, in the risk society, local and global merge and the personal becomes the political (Smith et al., 1997).

For Beck, environmental issues are at the forefront of subpolitical campaigns. The persistent lobbying across the globe by pressure groups such as Greenpeace and Friends of the Earth have succeeded in forcing risk onto the political agenda in the last 30 years. As will be elucidated in Chapter 4, the political case espoused by the Green movement has been reinforced by high-profile disasters and media reporting of environmental disasters. From the risk society perspective, environmental politics is a crucial terrain of political contestation. By way of illustration, Beck refers to the political conflict which resulted out of the proposed dumping of the Brent Spar oil rig by the Dutch company Shell (Beck, 1999: 40). Under substantial pressure from various non-governmental organisations and facing a decline in sales through consumer boycotting, the powerful multinational reneged on its decision to sink the oil platform in the North Sea. In this instance, methods of direct action effectively bypassed the formal system, highlighting the potential power of subpolitical protest. However, as we will learn in Chapter 8, subpolitics is much more than an arena where David is able to exact revenge on Goliath. Perversely, would-be enemies find themselves lining up on the same side of the table, albeit for dissimilar reasons. For example, in the debate over the labelling of genetically modified products, Greenpeace sided with food retailers Unilever and Iceland to lobby against the British government. As unlikely 'coalitions of opposites' participate in large and small-scale battles outside of the party process, a scattered set of environmental alliances emerges (Beck, 1998a: 7).

RISKS AND HAZARDS: A TENDENTIOUS DISTINCTION?

Having tracked the environmental slant of the risk society thesis, in the remainder of the chapter I wish to bring into play wider evidence in order to evaluate the association between risk, the environment and politics. By way of critique, the risk society perspective can be challenged on three fronts. First, the union forged between qualitatively distinct forms of risk and particular historical periods is contestable. Second, the empirical evidence marshalled to support a seismic shift in environmental conditions is deficient. Third, the actual impacts of environmental risk on routine practices are disputable, both in terms of everyday behaviour and political involvement.

The contention that western society can accurately be described as a risk society has proven to be the most hotly disputed aspect of Beck's work (see McMylor, 1996; Smith et al., 1997; Wynne, 1996). As described in Chapter 1, the movement into the risk society is dependent upon transmutations in the nature and impacts of risk. Whilst the realm of 'nature' has traditionally been distinguished from the realm of humanity, in the last half a century there has been mounting recognition that the environment should not and cannot be detached from human activity. Within academia, politics and science a more sophisticated appreciation of the interrelationship between nature and culture has evolved in the last three decades. In an attempt to reflect these developments, Beck counterposes the natural hazards common to pre-industrial cultures with the anthropogenic risks which envelop the risk society. However, under a modicum of pressure, this pillar of risk is easily toppled. The simple separation between natural hazards and manufactured risks is not suitably reflective of either the state of the art or the symbiotic relationship between nature and culture. Unsurprisingly, several critics have pointed up the pitfalls of linking discrete forms of risk to historical ages (Dryzek, 1995; Furedi, 1997; McMylor, 1996; Scott, 2000). In particular, controversy surrounds the supposition that contemporary risks can be productively separated out from the threats experienced in previous epochs. In as much as recognition of the connectivity between economic development and the lived environment may be a relatively modern phenomenon, human destruction of the environment is not. Environmental despoliation is a process which is centuries long, predating both the industrial and risk society markers laid down by Beck. Granted, the rate and frequency of risk production has accelerated in the last half a century in tandem with the intensification of capitalist production and consumption. Nevertheless, this does not warrant clumsy despatch of the period prior to 1950 as a 'pre-industrial society', exclusively blighted by natural hazards. In the first instance, it is not productive to allude to an amorphous pre-industrial period stretching out over thousands of years. Secondly, it is questionable whether natural hazards were ever 'natural' in any meaningful sense. Despite Beck's insistence that the risk society denotes a unique phase of techno-environmental hybridity, it is clear that nature and culture have always been intermeshed. Indeed, even the paradigmatic natural hazards of pre-industrial society – earthquakes, drought and flooding – cannot be freely divested from socio-natural processes (Hinchcliffe, 2000: 124). For example, we now know that floods are encouraged and aggravated by specific human practices. Ceteris paribus, uprooting of trees and foliage, removal of hedgerows, construction of sewerage systems and urban design all increase the velocity of water flowing into rivers and oceans. Similarly, although earthquakes are technically caused by the movement of tectonic plates, seismologists agree that the massification of buildings and the intensification of human movement have adversely affected the stability of the earth's crust. Although these qualifications may smack of captiousness, they do point towards a significant shortcoming in the risk society timeline and one which extends beyond a terminological quibble. In effect, the historical longevity of the interrelationship between the natural and the social questions the validity of the epochal distinctions made by Beck. Many 'manufactured' risks have 'natural' components and, conversely, natural hazards bear social imprints. Once we recognise this fluidity, the attribution of contrasting types of danger to different epochs is useful only as heuristic. Within the pre-industrial frame set by Beck we can identify any number of examples of anthropogenic risks. In ascribing 'natural hazards' to earlier epochs, the risk society argument overstates the boundaries between natural and social processes and understates human complicity in hazard production. If the risk society truly does begin 'where nature ends', it emerges somewhat sooner than Beck declares. Of course, banks of scientific knowledge are richer and deeper in contemporary western society than in pre-industrial cultures. Nevertheless, the human hand at play in the making of environmental dangers is more historically embedded than is accepted in the risk society thesis.

It is also the case that Beck accentuates the impact of manufactured risks and disguises the force of natural hazards. It must be remembered that – until they materialise – anthropogenic risks remain hypothetical. Moreover, although humanly designed nuclear and genetic risks are potentially catastrophic, Beck underplays the devastating environmental consequences of 'natural' disasters which have materialised in the past. For instance, in 1883 the tsunamis resulting from the volcanic explosion on Krakatoa caused 36,417 deaths (T. Green, 2003: 37). The eruption also produced global effects, with the shockwaves from the eruption travelling around the earth seven times and lasting for over two weeks (Winchester, 2003). Similarly, the volcanic explosion which wiped out the Minoan population is thought to have put as much matter into the atmosphere as the entire process of western industrialism (Leiss, 2000). Flying in the face of Beck's attribution of extraordinary geographical and catastrophic powers to contemporary threats, it is evident that a fair amount of convergence occurs between the properties of natural hazards and manufactured risks. Taken collectively, these qualifications and criticisms have the effect of questioning the structural solidity of the temporal and geographic pillars of risk.

ASSESSING ENVIRONMENTAL DAMAGE

In order to evaluate the contribution made to debates about the environment by the risk society perspective we must consider the extent to which its key claims are borne out by available evidence. As we shall see, certain aspects of the risk society argument hold water, whilst others are prone to leakage. Beck's treatise is undoubtedly strong in the areas of institutional critique and Green politics. However, the risk society thesis does not catalogue the full range of processes and practices that cause environmental disequilibrium. Nor is scientific evidence about the extent of environmental risk debated in any detail. To separate the insightful from the short sighted, it is worth comparing the risk society thesis with current scientific and academic knowledge about the environment.

As a shot in the arm for the risk society thesis, a growing body of research indicates that human activities have created wide-ranging environmental problems. The weight of evidence suggests that the climate has changed markedly over the last 300 years due to mass production and the burning of fossil fuels, such as gas, coal and petrol (Kamppinen and Wilenius, 2001: 312).[2] Most scientific experts now accept that climate change is induced by the escalation of greenhouse gases produced by the capitalist-industrial system (IPCC Report, 2001).[3] The 'greenhouse effect' – a rise in the earth's temperature due to increased production of carbon in the atmosphere – is taken as a given by most climatologists (Wilson, 2000: 203). Since the industrial revolution, it is estimated that between five and ten times the amount of carbon dioxide has been introduced into the atmosphere. Staggeringly, the global rate of carbon emissions is rising by around 2 per cent each year (Macionis and Plummer, 1997: 647). There is persuasive evidence of a mounting problem of global warming, with the earth's surface steadily heating up. Monthly and annual average temperatures have consistently increased over the last 100 years, with rises in temperature between 1900 and 2000 being the greatest during any century in the last 1,000 years (McCarthy, 2003b: 1). The World Meteorological Organisation – a United Nations-sponsored body – has used supercomputer models to demonstrate that extreme weather in different parts of the world are consistent with climate changes produced by global warming. Alarmingly, increases in temperature since 1976 are approximately three times the average for the 143-year period since 1860, with the ten hottest years on record all having occurred since 1990 (McCarthy, 2003b: 1).

In support of Beck, there is also a rich vein of evidence to suggest that capitalist methods of manufacture and consumption are steadily exhausting the earth's natural resources (see Hinchcliffe, 2000). The world's forests are now less than half their original size and are continuing to shrink by 65,000 square miles every year (Macionis and Plummer, 1997: 655). Last year alone, in the Amazonian rainforest an area the size of Belgium was destroyed (McCarthy, 2003a: 1). Further, the widespread destruction of plant life has a domino effect on the ecosystem. Because plants and trees consume carbon dioxide and exhume oxygen which restores the ozone layer, the destruction of plant life leads to increases in carbon dioxide and decreases the amount of oxygen in the atmosphere.

Today, climatologists agree that global warming is happening at a much quicker rate than previously envisaged, with temperatures expected to increase by anything from 1.4 to 5.8 degrees Celsius over the course of the twenty-first century (IPCC Report, 2001). As a consequence of accelerated melting of the polar ice caps, sea levels are expected to rise by tens of centimetres, increasing the threat of flooding in low-lying regions. In contemporary western societies, scientists who dispute the existence of global warming, ozone depletion and the exhaustion of natural resources find themselves in a minority: 'The evidence is mounting that, in our pursuit of material affluence, humanity is running up an environmental deficit, a situation in which our relationship to the environment, while yielding short-term benefits, will have profound, negative long term consequences' (Macionis and Plummer, 1997: 641).

From an everyday perspective, the detrimental impacts of manufactured risks on the lived environment are becoming more and more visible. In cities around the globe, breathing masks are routinely donned to combat air pollution. Flooding is becoming an almost routine occurrence in many areas. Meanwhile, the European agricultural industry has been blighted by a series of risk incidents, such as foot-and-mouth disease and BSE. All of this provides grist for the risk society mill. On the balance of evidence, Beck's 'eco-alarmism' is far from unfounded. While warnings about the potentially catastrophic consequences of capitalist expansion were seen as exaggerated 40 years ago, the injurious impacts of human actions on the environment are now accepted by scientific experts and policy makers within power-loaded institutional spaces (UNEP Report, 2002).

However, despite being scientifically justifiable and reflecting popular concerns about the environment there are a number of undeveloped areas in Beck's argument. Firstly, the risk society thesis does not work across the range of environmental risks and is haphazard in its coverage of environmental effects. Instead of providing a holistic analysis of the causes of environmental damage, Beck mixes repetitive references to the icons of destruction with a trek through the minutiae of particular cases. Insofar as this makes for entertaining reading, it is not conducive to tight theory building. Consequently, critics of the risk society perspective have highlighted Beck's speckled employment of scientific evidence and selective use of environmental examples (see Dickens, 1996; Goldblatt, 1995: 187).

These limitations are symptomatic of a broader tendency to assume that the effects of a small cluster of risks can be taken as representative of the totality of threats to the environment. The environmental mantra chanted by Beck gives voice to nuclear, chemical and genetic risks, global warming and air pollution. However, the overall impact of risks on the environment cannot be generalised on the basis of a 'highly selective account of the mismanagement of a few technologically induced hazards' (Leiss, 2000: 2). Considerable compositional disparities exist along a continuum of environmental threats (Mol and Spargaaren, 1993: 431). For instance, the effects of global warming are vastly different from those attributed to nuclear power. Furthermore, not all 'risks' to the environment will have detrimental consequences in the final analysis. Even though Beck's icons of destruction are undoubtedly anxiety-provoking, it does not follow that they will indubitably result in global catastrophe. In certain circumstances, the trade-off between what might be lost, as against what can be gained, is a matter of opinion (see Philmore and Moffatt, 2000: 111). Although nobody would sensibly argue that chemical pollution is socially productive, the manufacture of chemical substances has improved methods of food production and enhanced the quality of medicinal treatments (Smith, 2001: 148).

If truth be told, the risk society representation of environmental hazards is far from impartial. As Goldblatt points out, Beck is unwilling to engage with ideas which contradict the party line: 'It is almost as if Beck assumes that we agree with his estimations of the dangers we face – a somewhat surprising elision given Beck's acknowledgement of the relative and contested character of risk perception and definition' (Goldblatt, 1995: 158).

Notably, the risk society perspective fails to confront conservative and/or neo-liberal philosophies which conceive of risk as a precursor to social change and prosperity. It should be recognised that some dyed-in-the-wool traditionalists deny that the environment is irreparably damaged. A small but vociferous group of scientists, economists and politicians still maintain that global warming is part of a natural cycle. Others claim that the earth's climate is party to routine fluctuation, pointing out that the planet was substantially warmer 6,000 years ago than it is today (see Wilson, 2000: 202).[4] On the margins, heterodox scientists have contended that higher levels of carbon dioxide in the atmosphere actually encourage plant growth (Macionis and Plummer, 1997: 655). Although now is not the moment to unwrap the science of such arguments, it is worth remarking that

Beck's representation of environmental issues is decidedly one-sided and bypasses oppositional opinions (North, 1997). In his haste to scold apologists of unfettered economic development, Beck forgets to counter ideological challenges to the Green agenda. Instead of taking establishment science to task on the issues, the risk society thesis settles back into amplifying the message that institutionally generated techno-scientific risks make real the prospect of planetary annihilation.

In many respects, Beck's absorption with the risk-producing tendencies of capitalist institutions obscures the multi-agential production of environmental risk. While capitalist institutions play the villains of the environmental piece, other culpable parties recede into the background. It ought to be mentioned that an assortment of factors have caused ecological disequilibrium, including dramatic population growth and spiralling patterns of consumption. In 1800 a billion people drew on the Earth's resources. By the year 2000 the figure had increased sixfold. Staggeringly, it is estimated that those born in 1950 will have witnessed more population growth in a lifetime than has previously occurred during the rest of human history.[5] The implications of this demographic explosion for the lifespan of finite resources are self-evident.

In *Ecological Politics in an Age of Risk* (1995a) Beck launches a scathing attack on the excesses of the capitalist system. Science, business, law and government are all reproved for creating a culture of rampant insecurity. Yet Beck is less strident about the negative effects of individual consumer choices in capitalist cultures. Superfluous consumption, disposal of renewable materials, the wastage of finite energy sources and emissions of carbon monoxide and CFCs all produce damage to the environment. In the twentieth century alone, the per-head level of consumption in the United States of America grew by ten times.[6] Every day in Britain the equivalent of five football stadiums' worth of waste material is deposited in landfill sites. By 2020 it is anticipated that the rate of personal waste produced will have more than doubled.[7] This is not to posit that the burden of responsibility for environmental demise should be shared between the general public and capitalist institutions. Clearly, risk-regulating institutions have the material resources, the legal jurisdiction and the political authority to make policy in power-bound spaces. Nevertheless, as will be elaborated in Chapter 3, the risk society perspective conceals public responsibility for environmental despoliation so as to accentuate institutional culpability.

ENVIRONMENTAL AWARENESS AND POLITICAL ACTION:
A CRITICAL DISJUNCTURE?

Prior to concluding, it is worth returning to the association established between environmental harm and political engagement in the risk society thesis. As outlined earlier, Beck alludes to a positive correlation between the perceptibility of manufactured risks and lay awareness of environmental issues. This is crystallised in public disenchantment with expert institutions funnelled through the activities of environmental pressure groups. The Green movement has unquestionably grown in size and influence since the first wave of environmentalism stimulated by the pronouncements of the 'Club of Rome' in the 1970s. The stark warnings issued by the group catapulted the problem of environmental sustainability onto the political agenda and heightened public sensitivity towards environmental issues (see Mol and Spaargaren, 1993: 436). A decade on, a second wave of environmentalism emerged in the 1980s, with Green campaign groups spearheading calls for institutional recognition of the interconnected nature of the ecosystem. Along with Beck, it would seem reasonable to suggest that the general state of environmental awareness in contemporary western cultures is relatively advanced. There is greater social realisation of the relationship between ecological processes and human decisions than at any other historical juncture. In the 1960s and 1970s environmental ideology collided with the values of the establishment and formed part of a broader countercultural movement. By the new millennium, most individuals employed in social institutions recognise rather than rebut environmental problems. Global warming, ozone depletion and the greenhouse effect have all become part of political, scientific and public discourse over the last three decades. Meanwhile, general interest in environmental issues is reflected in the numbers joining environmental groups, such as Greenpeace and Friends of the Earth.[8]

Obviously, there are a number of reasons for increased public apprehension about environmental issues. Scientific inquiries have reported damage to the ecosystem, national and local governments have launched Green awareness campaigns and environmental issues have been helped up the ladder by the agitations of new social movements. Direct action, entry into power-rich political spaces and sophisticated media management have raised the profile of the Green movement (Strydom, 2003). Environmental pressure groups have successfully diversified their activities, lobbying with and against a diverse range of interest groups, from national governments to electricity suppliers.

In a hostile political age, it is unsurprising that the risk society perspective has struck a chord with ecologists keen to advertise environmental issues. Nonetheless, stepping back from partisanship and returning to the task of academic scrutiny, it is clear that Beck's reliance on anecdotal examples gives rise to an incomplete analysis of the association between environmental risk and political motivations. We might reasonably request the evidence of a linear link between risk, behavioural change and political activity. Empirical studies do suggest a gross increase in public awareness of environmental risks over the last quarter of a century (see Macnaghtan, 2003; ONS, 2001). Nevertheless, the net effect of information about environmental risk on everyday behaviour and political motivations remains difficult to estimate. At the very least, it would seem rash to presume that awareness of environmental issues is being translated into behaviour modification. Without foraging for evidence, Beck makes a bold leap from cognition of environmental risk to changing values and practices. However, the distance between cognition and enactment proves unjumpable. As has long been established within critical social theory, consciousness cannot simply be equated with action (Lodziak, 1995: 40; Mann, 1982: 388). The so-called 'value–action gap' indicates that recognition of environmental issues does not readily translate into ecologically sound behaviour (Blake, 1999, Munton, 1997). Rather, there are a series of intervening factors which obstruct the conversion of environmental awareness into greener practices. Doubtless, many people would acknowledge that driving to work is an environmentally unsound activity.[9] Nevertheless, mitigating factors – the spacing out of home from work, ineffectual public transport and the need to ferry children to school – may intercede. Running with the ebb and flow of everyday life, the need to adhere to long working hours and customary family commitments may short the circuit between best intentions and actual practices. As a result, many people will choose to live with fleeting ethical scruples, rather than embark on the difficult and time-consuming process of restructuring daily schedules. This is a vital point, for at least two reasons. First, tangible improvements in the quality of the environment depend upon significant cultural changes in employment patterns, production processes, consumer behaviour, transport provision and energy consumption. Without radically altering the way we live, work and move in capitalist cultures, a reversal of the process of environmental despoliation looks remote. Second, as will be expanded upon in Chapter 8, the emancipatory capacity which Beck attaches to the risk society very much depends upon the durability of the link between risk consciousness and political action.

In contrast to the totalising approach of the risk society thesis, the degree of commitment to environmental restoration cannot be generalised. Beck envisages a highly educated, ecologically schooled and politicised public in possession of the economic capacity to choose environmentally friendly options. In reality, the preferences of such a privileged group cannot possibly be reflective of all publics. In the affluent West, interest in and knowledge about environmental affairs is inextricably tied to education, income and life chances. Branching out, the luxury of making Green choices does not extend into vast tracts of Africa, Asia and South America. For poorer nations, attention to long-term planetary effects may not be high on a wish list topped by nourishing the population, providing basic health facilities and ensuring adequate sanitation levels. What is more, the degree of complexity associated with environmental risks directs the extent to which they are amenable to explanation. Suffice it to say that scientific models and discourses cannot be magically translated into clear and understandable language (Allan et al., 2000: 13). Although public recognition of environmental issues has risen in general, this cannot be passed off as evidence of extensive knowledge of environmental affairs across different groups. The reality of the situation is likely to be shades of grey, not black and white. Strands of research to be revealed in Chapter 5 have highlighted relative sophistication in lay understandings of environmental risks (ESRC Report, 1999; Reilly, 1998), However, other studies report significant contradictions and confusions. Wilson (2000: 209), for example, found that many people conflate global warming and the greenhouse effect, instead of seeing global warming as one effect of increased greenhouse gas concentrations in the atmosphere. Passing over the ins and outs, public knowledge of environmental affairs is likely to be variable and relative rather than absolute. One may have a firm grasp of the process of global warming, but be perplexed by the intricacies of the debate about genetically modified organisms. Empirical studies have shown that lay models of the connections between different parts of the ecosystem can be fuzzy and are characterised by ambiguous and/or contradictory understandings (see Kamppinen and Wilenius, 2001: 316; Lazo et al., 2000). Although most people are aware that interrelationships exist between environmental elements, the precise links between different environmental processes are not always clearly comprehended or articulated.

Returning to Beck's work, public knowledge of the interconnectivity between human actions and environmental risks cannot be assumed en bloc. Although it would be fair to say that a greater number of people in the West have become more aware of a wider range of environmental issues, it cannot be deduced that knowledge is evenly spread or coherently organised. Nor does it suggest that people are able or willing to convert to ecologically sound practices. Some people will be knowledgeable and interested in environmental affairs, others will have knowledge gaps or conflicting priorities. Whatever the spread, the vast majority will be bound up with attending to the immediate demands of work, family life and personal relationships. Allowing for the weighty burden of the realm of necessity, the abstract prospect of 'environmental risk' may not make as much headway as Beck supposes. Against the risk society logic, people may treat environmental risk not as a coherent and omnipresent issue, but as and when it connects with their everyday experiences. Instead of relating to a single conception of 'the environment', people may work with different environments, according to their beliefs, pursuits, motivations and aspirations (Macnaghtan, 2003).[10]

THE FRAGILITY OF ENVIRONMENTAL TRUTH CLAIMS

Most theorists would agree that the profile of environmental pressure groups has risen in tandem with social recognition of the production of detrimental risks (Anderson 2000: 95; Goldblatt, 1995). In the course of successful campaigns, pressure groups such as Greenpeace have forced political power holders to address particular environmental issues. However, regardless of whether the public are generally supportive of the principles of the Green movement,[11] it should not be assumed that pressure groups are the arbiters of environmental truth. The risk society thesis clumsily assumes that Green knowledge about risks is uniformly superior to information disseminated by state institutions. This is misleading, when one considers that agencies such as Greenpeace cannot possibly insulate themselves against imperfect information, particularly given their commitment to swift and direct action. This point is vividly illustrated by the Greenpeace–Shell Oil dispute in 1995. After discovering that Shell was intending to sink the Brent Spar oil platform in the North Sea, Greenpeace instigated a series of direct protests and encouraged the public to boycott Shell products. The major bone of contention was the quantity of hydrocarbons on the Brent Spar oil platform, which Greenpeace described as 'dangerously high'. Disputing this summation, both Shell and the British government claimed that the level of hydrocarbons on the platform was negligible and that the sinking of the platform constituted an infinitesimal risk to public health. Eventually, Greenpeace activists occupied the oil platform and succeeded in preventing its submergence.

On paper this example serves as an exemplar of effective subpolitical protest driven by eco-social rationality. However, with respect to the competing truth claims about risk, the story has a distinct twist in the tail. As Anderson explains:

> While Greenpeace estimated that the amount of hydrocarbons was in the region of 5,000 tonnes, an independent investigation by the Norwegian certification company, DNV (Det Norske Veritas), revealed there were only around 75–100 tonnes of oil on board. Shell's original estimate of around 50 tonnes of oil thus proved to be considerably closer to the final figure. Greenpeace wrote to Shell apologizing for the error. (Anderson, 1997: 112)

The Greenpeace–Shell dispute articulates the difficulties which arise in assuming that truth about environmental risk 'belongs' to certain organisations and not to others. Given the uncertainty associated with manufactured risks, Beck's attachment of verity to the public and environmental groups would appear to be misguided. Somewhat ironically, having documented that risk-defining institutions silence diverse environmental voices, Beck proceeds to assume an omniscient stance on the meaning and the effects of risk (Alexander and Smith, 1996: 254). In this way, the assumptions of the Green movement are rendered as objective facts, rather than vigorously investigated and evaluated. While Beck should not be denigrated for assuming a political viewpoint, if we are to arrive at a balanced understanding of the impacts of environmental risks, it is essential that both sides of the debate are at least aired. Simply replacing one absolute with another seems a peculiar form of environmental democracy. Although Beck's political ambitions are unquestionably worthy, the risk society thesis would be enriched by greater sensitivity to the partiality of all environmental truth claims, not only those which flow from dominant institutions. In times of social and scientific indeterminacy, neither academics, citizens, experts or environmental groups can hope to speak with complete certainty about risk.

CONCLUSION

In conclusion, it is clear that the risk society perspective adds to our understanding of the lived environment in a number of ways. First, Beck's thesis addresses one of the most pressing and socially significant problems of the modern age. Whether or not there has objectively been an increase in the range of environmental dangers is a difficult question to answer, with a great deal depending upon the point of comparison. Insofar as scientific technologies have allowed us to 'uncover' environmental threats, many of these dangers have been in progress for many years. Taking a macro view, it is probable that environmental risks have increased, both in terms of their geographical range and the scale of their potential effects. Frankly put, if we do not alter the course of capitalist development, the longevity of the planet will be curtailed:

> The current profile of environmental dangers looks more risky than, say, that of forty years ago: the ozone holes are bigger; the apparent trends of climate change are more ominous; more land is under threat from infrastructures and industrial development or desertification; there are innumerably more nuclear installations dotted across the earth; the state of the world's oceans is declining. (Goldblatt, 1995: 175)

Second, the risk society thesis deftly matches up the various economic, political and scientific parties involved in the production and management of environmental risks. Beck's account of the systemic failure of risk-regulating institutions is gripping, dense and provocative. At a practical level, the risk society argument highlights significant flaws in institutional procedures within law, politics and science. It is quite possible that reference to Beck's work in high-profile documents has served to fix political attention on the inadequacy of existing environmental regulations (CSEC Report, 1997; ESRC Report, 1999).

On the minus side, given his undoubted knowledge of environmental affairs, it is surprising that Beck subscribes to a schema which endorses a purified view of nature. First, the attachment of different forms of risk to historical periods is not sensitive to anomalies, nor is it receptive to historical and cultural difference. On the evidence presented here, it is clear that the separation between natural hazards and manufactured risks is wanting. Environmental risks are more diffuse and complex than can be accommodated in the risk society framework. Significant overlaps emerge between types of risk over time, space and place (Smith, 2001: 160).

Second, we have questioned the extent to which Beck presents a balanced description of the role of social institutions in generating and regulating environmental risk. As will be explicated in Chapter 3, the risk society thesis forces various institutional attempts to manage environmental risk into a single box marked 'obfuscation'. It is progressive and productive to criticise particular environmental policies, but it cannot be ahistorically imagined that *all* institutional power brokers in *all* nation states have sought to mask environmental damage. In different epochs and over different cultures institutional responses to risk have varied considerably, as has the efficacy of environmental initiatives.

Third, the risk society thesis presents an unrealistically optimistic picture of public commitment to environmental politics. Although the detrimental effects of environmental risks may be recognised by the majority in the West, the shadow of catastrophe has had a more limited impact upon customs and practices. In terms of the movement toward a more ecologically responsive culture, tangible changes in environmental values and socio-economic practices are required. Ironically, it is the economically, technological and culturally powerful nations who continue to generate the highest levels of harmful emissions.[12]

Finally, we should be concerned that the geographical spread of 'global risks' is conspicuously patchy. Certain risks will occur more frequently and more intensely in some regions than others (see Bromley, 2000). Despite Beck's insistence that environmental risks pose a global threat, we need to commit to memory that ecological hazards do not occur by happenstance and are not randomly scattered. Continents such as Africa, South America and Asia are party to a disproportionate range of environmental hazards and may act as convenient risk repositories for the West (Smith and Goldblatt, 2000: 101). In short, 'vulnerability to disaster is an uneven matter' (Hinchcliffe, 2000: 131). Quite naturally, the idea of a universal axis of risk distribution appeals to the egalitarian principles of western academics. However, it may be less palatable to those living against the hard edge of environmental risk in Beijing and Sao Paulo. For these global citizens, the suggestion that 'smog is democratic' may prove a little hard to swallow.

3
Defining Risk

Having considered the systemic production of environmental dangers, I now wish to zoom in on the process of risk definition. In particular, we will be interested in the role of social institutions in rendering risk calculable and meaningful. To this end, the interrelationship between science, law and government will act as a fulcrum for debate. Again, we begin by presenting the bones of Beck's argument, outlining the historical emergence of economic and scientific methods of risk identification and contrasting modes of definition in industrial and risk societies. From here, we move on to explore the crisis of knowledge which emerges in the risk society, setting down the battle between scientific and social rationality. In an attempt to ground abstract theory, we use the BSE imbroglio as a means of shedding preliminary light on the changing relationship between scientific experts and the general public.[1] On the basis of these wings of inquiry, I gather together and develop the principal objections to Beck's account of risk definition and evaluate the explanatory power of the risk society perspective. Providing a glimpse of the issues to be considered in Chapter 4, we will also begin to roughly cultivate the land between the definition, representation and mediation of risk.

THE HISTORY OF RISK DEFINITION

In order to subject the risk society narrative to critical scrutiny, it is necessary to offer a historical account of the institutional dimensions of risk definition. Tracing the blueprint set by the risk society thesis, in this section we consider the extent to which dominant institutions involved in risk assessment have lost potency in recent years. As discussed in Chapter 1, the collective relations of definition are seen as primary definers, organisers and regulators of risk. In turn, institutional definitions of risk inform and influence public interpretations (Beck, 1995a: 130). In a number of texts (1992; 1995a; 1999; 2000) and a raft of journal articles (1987; 1992; 1996a), Beck has reviewed the ways and means by which dominant institutions have historically generated risk meanings in western cultures. The central theoretical feature of the risk society take on the construction of danger is the notion of 'relations of definition'. According to Beck: 'The relations of definition are ... basic principles underlying industrial production, law, science, opportunities for the public and for policy. Relations of definition thus decide about data, knowledge, proofs, culprits and compensation' (Beck, 1995a: 130).

Although the term is notoriously slippery, the relations of definition are best understood as a panoply of institutions and agencies involved in the uncovering and communication of risk (Wales and Mythen, 2002). Beck considers government, science, law and the mass media as the institutions strategically implicated in hazard detection (Beck, 1995a: 61).[2] However, the broader spheres of science, business and technology also make up the relations of definition. Allied to their responsibilities for identification, the relations of definition are also charged with risk assessment and risk regulation. For example, national governments, scientific experts and legal professionals are involved in determining levels of risk acceptability *and* deciding upon appropriate compensation packages in cases of harm.

While the risk society critique is attentive to the operations of a range of institutions, science is presented as the driving force behind risk definition. For Beck, the principles and practices of scientific inquiry have historically acted as a keystone for social development. In the Enlightenment period, science was seen as the mechanism through which society could achieve mastery over nature. While the risk society critique is attentive to the activities of a range of institutions, science is presented as the historical driving force behind risk definition. For Beck, the principles and practices of scientific inquiry have acted as a keystone for social development. In the Enlightenment period, science was seen as the mechanism through which society could achieve mastery over nature. By the end of the nineteenth century, various breakthroughs – such as Darwin's theory of evolution, Edison's work with electricity and Pasteur's germ theory of disease – had amplified the ideological momentum of science. Following on from this period, scientific and technical discourse came to dictate discussions about risk, which were formulated within the parameters of economics and the natural sciences (Beck, 1992: 24). Post-Enlightenment, dominant methods of explaining risk moved away from religious fate and towards technical and scientific rationality. This transition was reflected in the nineteenth century by the emergence of a distinct institutional form, whose *raison d'être* was to protect citizens from potential dangers. Borrowing from the work of

François Ewald (1986; 1991), Beck (1995a: 108) refers to this body as the 'safety state'. In the risk society narrative, the ideology of the safety state is instrumental in developing cohesion between the various components of the relations of definition (Beck, 1995a: 116).

Imitating technical and scientific rationality, the relations of definition in capitalist societies were constructed on a calculative basis (Beck, 1995a: 85). The model of risk attribution cultivated by economics and the natural sciences is referred to as the 'calculus of risk' (Beck, 1995a: 77). The calculus of risk is essentially an economic paradigm, through which methods of risk assessment are tied to the principles of mathematics and probability. By way of example, actuarial insurance systems arose to determine the probability of accident occurrence, national legal systems were introduced to determine liability and the welfare state evolved to improve public health. The supremacy of the calculus of risk in early capitalist societies ensured that hazards were quantifiable:

> Regardless of the size of a workforce or the turnover of its recruits, a given mine or factory will show a consistent percentage of injuries and deaths. When put in the context of a population, the accident which is taken on its own to be random and unavoidable, can (given a little prudence) be treated as predictable and calculable. One can predict that during the next year there will be a certain number of accidents, the only unknown being who will have an accident, who will draw one of existence's unlucky numbers. (Ewald, 1991: 202)

Following Ewald, Beck asserts that models of insurance in industrial society were based on actuarial principles by which the probability of an accident could be calculated. In the period from the early nineteenth century to the mid twentieth century, the rules and regulations constructed by the relations of definition were geared towards handling tangible and attributable risks. Through institutionally ingrained methods, sources of risk were recognised, guilty parties punished and compensation packages awarded to victims. At this historical moment, the accumulated body of knowledge about risks, allied to the rules and regulations limiting harm, enabled the welfare state to foster a climate of relative security for its citizens (Beck, 1995a: 85).

Obviously, the dominance of the calculus of risk in the nineteenth and twentieth centuries can be fastened to the expansion of economic and scientific values. Post-Enlightenment, notions of human development were bound up with advances in the economy, science, technology and medicine (Polanyi, 1975). Beck believes that public trust in techno-scientific development can be traced from the Enlightenment period through early industrial society and onwards into the advanced stages of industrial capitalism in the mid twentieth century (Beck, 1992: 155). In the 1950s, the use of assorted machinery in the production process, the manufacture of health vaccines and developments in the space race reinforced public faith in science and technology as motors of social progress. Although we will return to question the veracity of this storyline, for now, we move on to weigh up the activities of the relations of definition in industrial and risk societies.

THE EMERGENCE OF CONFLICTING RATIONALITIES

In contrast to pre-industrial times, in contemporary society risks are primarily constructed through the relations of definition rather than by recourse to mystical beliefs or religious ideology (Beck, 1992: 27). In the twentieth century, science presided over matters of risk iden-tification, acting as a tool for investigation and empowerment. By dint of this, scientific experts have traditionally been cast as the talking heads through which environmental risks are articulated to the public. Supported by the safety net provided by a universal methodology, scientific and governmental experts steered not only the process of risk definition, but also the formative content of debate about risks. Having identified science as the organising mode of inquiry within the relations of definition, Beck moves on to raise a series of issues about the relationship between the lay public and scientific experts. In simple industrial society, the dominant direction of information about risk flows from expert assessors to the lay public. This imbalance of communicative power is padded out with reference to the competing values of 'scientific' and 'social' rationality (Beck, 1992: 29). Scientific rationality refers to dominant technical discourses utilised by scientific experts. Conversely, social rationality stems from cultural evaluations convened through everyday lived experience. During the industrial phase, scientific ideals are trusted and valued, with technological advancement perceived as a route to prosperity. In the consensual climate of industrial society, the distance between social and scientific rationality is inconsequential. However, in the transition from industrial society to the risk society, manufactured threats repeatedly pop up, bursting institutional boundaries and driving the two rationalities apart. As lay actors become ever more frustrated by the ineptitude of the relations of definition, conflicting values and objectives become discernible. To twist Thomas Kuhn's (1970) memorable phrase, a case of 'paradigm incommensurability' exists:

> The two sides talk past each other. Social movements raise questions that are not answered by the risk technicians at all, and the technicians answer questions which miss the point of what was really asked and what feeds public anxiety. (Beck, 1992: 30)

As exemplified by the environmental disputes considered in Chapter 2, the risk society is distinguished by an ongoing conflict of meaning between experts following the guidelines of scientific rationality and the lay public gazing through the lens of social rationality. In line with the development of a broader environmental critique, Beck is disparaging toward expert bodies, particularly those that have prioritised scientific over social rationality.[3] To dramatise the failure of institutional forms of regulation routine references to the icons of destruction are mixed with details of high-profile risk incidents, such as Chernobyl and Bhopal.

In the risk society thesis, the value conflict between social and scientific rationality can be traced back to the early workings of the safety state. As has been noted, the calculus of risk became the stock method of calibration from the eighteenth through to the mid twentieth century. However, in the mid to late twentieth century, the legitimacy of the calculus of risk becomes threatened by the generation of unmanageable risks which began to outstrip prevailing methods of calculation and liability:

> Studies of reactor safety restrict themselves to the estimation of certain *quantifiable* risks on the basis of *probable* accidents ... in some circles it is said that risks which are not yet technically manageable do not exist – at least not in scientific calculation or jurisdictional judgement. (Beck, 1992: 29)

By the dawn of the risk society, it is apparent that the environmental side effects of potent nuclear, chemical and genetic technologies cannot be adequately regulated or managed through existing channels of risk assessment. The very institutions charged with containing hazards are incapable of limiting the production of manufactured risks.

Through pointed critique it is argued that scientific experts have traditionally been guilty of hoarding expertise about risk and overlooking the usefulness of social rationality. Beck claims that, amongst some scientists, public conduct which deviates from expert advice has been interpreted as evidence of miscomprehension. The logical – but ultimately divisive – remedy is the provision of more 'hard facts' about risk. However, the risk society thesis goes much further than simply hailing the seeds of the conflict between social and scientific rationality. The radical break away from industrial society and towards the risk society is symptomatic of broader changes in the relationship between institutional experts and the general public. In the risk society, dangers become multiple and expertise becomes contingent rather than fixed. As environmental dangers ooze into the popular consciousness, active public engagement with risk issues promotes a 'scientised' consciousness (Beck, 1992: 28). As we shall see in Chapter 4, this reflexive public consciousness is informed and nurtured by the mass media which throws the objectivity and purpose of science into question. In the risk society, the growth of the media and the diversification of information lead to the ideological unbuttoning of social institutions. Through perpetual concentration on uncertain and risky situations, contra-dictory theories emerge: 'if three scientists get together, fifteen opinions clash' (Beck, 1992: 167). Indeterminate risks stimulate a rash of competing truth claims and expert constructions of truth become pluralised. This point is aptly illustrated by the heated speculation that has surrounded the likely cause of BSE in cattle (Ratzan, 1998; Reilly, 1999; Wylie, 1998). The diversification of scientific opinion is eagerly seized upon by the media as evidence of expert disagreement and endemic uncertainty. In the risk society, media scrutiny encourages the public to catechise expert discourses of risk (Beck, 1987: 158). *Par consequence*, unconditional public trust in science has all but evaporated: 'the true–false positivism of clear-cut factualist science, at once this century's article of faith and its terrifying spectre, is at an end' (Beck, 1995a: 119). As public reflexivity blossoms, protest groups and counter experts begin to delve deeper into the relationship between science, politics and business. Each and every revelation of collusion between the parties of definition destabilises the authority of scientific rationality: 'the very institution which has been employed as a tool to disenchant religious and fatalistic beliefs, itself, in turn, becomes disenchanted' (Beck, 1992: 256). Beck is striving then to force home a broader point about shifting public conceptions of science and technology. In the risk society, unbroken techno-scientific development serves to *produce* rather than eliminate risks to health. Hence, social institutions – hitherto extolled as the saviours of society – are recast as the progenitors of risk.

Because there is no definitive authority on risk, public belief in scientific rationality erodes and lay actors themselves become 'small, private alternative experts in the risks of modernisation' (Beck, 1992: 61). In such conditions of 'reflexive scientisation', the mobilisation of beliefs is pivotal to the success of competing truth claims about risk (Beck, 1992: 169). Reflexive scientisation empowers the public in their struggle for safety and equality, enabling lay actors to challenge dominant relations of definition: 'to speak up, organise, go to court, assert themselves, refuse to be diverted any longer' (Beck, 1992: 77). In such an adversarial social climate, public debate heightens political consciousness and stokes up contestation about environmental risks. Nonetheless, it is only through the application of science that solutions to risks – for example, vCJD or Aids – can be unearthed. Thus, the populace of the risk society are situated between a rock and a hard place. They are ultra-critical of science and government, whilst simultaneously being dependent on expert systems for solutions to manufactured uncertainties. Even an attitudinal revolution does not necessarily loosen the ideological grip of science and technology. Because threats to health can only be formally identified through scientific inquiry, the general public are ultimately reliant on science, regardless of whether risks are visible or not. Perversely, it is possible that individuals have become more rather than less dependent upon scientific and technical experts:

> That which impairs health or destroys nature is not recognizable to one's own feeling or eye, and even where it is seemingly in plain view, qualified expert judgement is still required to determine it 'objectively'... hazards in any case require the 'sensory organs' of science – *theories, experiments, measuring instruments – in order to become visible, or interpretable as hazards at all.* (Beck, 1992: 27, emphasis in original)

ORGANISED IRRESPONSIBILITY: PAPERING OVER THE CRACKS

As we have seen, in addition to being the self-proclaimed agents of risk management, the relations of definition are also hailed as manufacturers of risk. Science, government and technology are publicly acknowledged to be 'taboo constructors', as well as a 'taboo breakers' (Beck, 1992: 157). This paradoxical position of bearer and regulator places the relations of definition in something of a fix. Given that social institutions are reliant on public consent, it is in the short-term interests of government, law and science to conceal and deflect hazards (Beck, 1995a: 86). For the relations of definition, accepting responsibility for the manufacture of risks *and* admitting inability to contain hazards would lead to a crisis of legitimation. For Beck (1995a: 61), the institutional response is a diffuse process of 'organised irresponsibility'. In *Ecological Politics in an Age of Risk* (1995a) Beck relates the notion of 'organised irresponsibility' to the problem of risk regulation. The concept of organised irresponsibility is the result of a mismatch between contemporary risks and the safety capacities of the relations of definition. Organised irresponsibility refers to the way in which institutions are forced to recognise catastrophic risks whilst simultaneously refuting and deflecting public concerns: 'Thus what is at issue is an elaborate labyrinth designed according to principles, not of non-liability or irresponsibility, but of simultaneous liability and unaccountability: more precisely, liability as unaccountability, or organised irresponsibility' (Beck, 1995a: 61).

   To flesh out the concept of organised irresponsibility, Beck describes a phalanx of strategies utilised by the relations of definition, including denial, misinformation and mystification (Beck, 1995a: 64). It is argued that institutional strategies of risk diffusion involve the performance of 'symbolic detoxification' (1992: 65; 1995a: 84). Risks such as nuclear and genetic technology are 'detoxified' by defining institutions. The 'staging and perfecting of a cosmetic treatment of risks' (Beck, 1995a: 84) works its magic through a blend of reassurance and mystification. Symbolic detoxification is operationalised through the repetition of 'scientific evidence' and reassurance of future control (Adam and van Loon, 2000: 13). If successful, the ideological exercise of symbolic detoxification is detrimental to public health in two ways. Firstly, it is possible that sections of the public may be placed in danger by presuming dangerous activities to be safe. Secondly, the institutional concealment of hazards serves to foster the unchecked development of upcoming risks. This leads Beck to postulate that the relations of definition are fundamentally incompatible with the risks they are charged with containing (Beck, 1995a: 160). The volatile and unpredictable nature of manufactured risks allied to the impotence of existent institutional mechanisms means that the collective safety of citizens has been compromised. As manufactured risks grow, the apparatus of definition become increasingly defensive, forced into the cynical tactics of organised irresponsibility. In the risk society, public recognition of the complicity of science and technology in the production of dangers becomes widespread and the relations of definition suffer continual bouts of instability: 'science has just lost the truth – as a schoolboy loses his milk money' (Beck, 1992: 166). The crisis of confidence in the relations of definition is exacerbated as risks become 'socially exploded' by the mass media (Beck, 1995a: 96). The 'social explosiveness of hazards' means that social institutions are left forcing a finger into a dam that has long since crumbled.

BSE: THE RISK SOCIETY IN MOTION?

Having set up the risk society take on the institutional construction of risk it is now time to take a closer look at the concepts discussed. To what extent can the theoretical tools of Beck's trade be applied in practice? To render the abstract concrete, we employ the BSE crisis as an archetypical case of the public–expert struggles which pepper the risk society landscape.

Reflecting academic consensus, Beck (1998b: 9) believes that the BSE episode acts as a historical benchmark for public trust in social institutions. As the Phillips inquiry (2000) indicated, a chain of institutional errors were made in the detection, management and communication of risk. Indubitably, the concept of organised irresponsibility can be aptly related to governmental responses to the BSE crisis. In the late 1980s and early 1990s, the Conservative government in Britain consistently denied any link between BSE in cattle and a variant of the brain disease Creutzfeldt-Jakob syndrome in humans. As explained in the Phillips report (2000), for over a decade – and despite being furnished with a wealth of information indicating a possible threat to public health – the Cabinet chose to adopt a strategy of denial and dismissal. Following the first wave of public concern about the transmission of BSE from cattle to humans in the late 1980s, the government were fearful of a loss of consumer confidence and a subsequent fall in export profits. In response, a troupe of scientific experts were recruited to play down the possible connection between BSE and vCJD in humans (Harris and O'Shaughnessy, 1997). During one exceptionally distasteful incident, the Environment Minister John Gummer was filmed by the media feeding beefburgers to his four-year-old daughter, Cordelia. This symbolic gesture was, of course, a rather desperate attempt to allay fears about the risk of eating British beef and to inveigle the public into further consumption. As discussed by respondents in Reilly's (1999: 135) study, the outright denial of potential risks by senior politicians was both disingenuous and misleading. Given the range of scientific information available to government ministers at the time, to arrive at the conclusion that eating British beef presented no risk whatsoever to public health was either grossly incompetent or simply mendacious. In flatly refuting the existence of potential risks, Gummer and his colleagues were, in effect, lying to the British public. Buckling under an embarrassing weight of evidence, the Conservative government eventually admitted the link between BSE and vCJD in March 1996 (Reilly, 1999: 134). Since this time, cattle infected with BSE have been found across Europe in Germany, France and Spain. More alarmingly, mad cow disease has appeared in homegrown cattle in countries as far afield as Canada and Argentina.

Beck proposes that BSE and the polemic surrounding it illustrate the negative consequences of socially exploding hazards on trust relations between experts and the public. In such cases, the powerlessness of risk-regulating institutions – science, medicine, law and government – are writ large. In the wake of BSE, national governments have continued to slope shoulder the issue, reassuring the public with safety claims about a risk which is effectively out of their orbit of control. Since the risk of contracting vCJD through eating beef broke post consumption, legislation and government policy only function at a perfunctory level. As far as the science of the matter is concerned, there is little chance that current research will provide a reliable indicator of the level of risk to public health (Hinchcliffe, 2000: 142). As Barbara Adam notes, experts have yet to develop a robust method of testing for BSE or variant CJD:

To date a brain biopsy is the only means to confirm the presence of BSE. Its incubation period is variable and may or may not be dependent on the dose of infective material. There is certainty neither about the infective agent or how it enters the body. The hypothesised prion protein (PrP) as infective agent has not been empirically verified. Yet, it appears that the infectivity is resistant to heat, chemicals and radiation. The same uncertainty applies to vCJD, whose lesions in the human brain tissue are similar to those found in BSE infected cattle. (Adam, 2000b: 118)

In the aftermath of the BSE imbroglio, it is tempting to concur that expert systems are struggling to contain the force of manufactured risks. It does seem reasonable to argue that the combined actions of social institutions have not succeeded in ensuring a safe society. In questioning the methods of risk assessment deployed by dominant institutions, Beck finds himself in accord with the public mood. In recent times, scientific rationality has certainly been out of tune with public expectations (see Douglas, 1985; Thompson, 1989; Wynne, 1989; 1996). Due to its linear understanding of the relationship between risk, perception and behaviour, the scientific model has erroneously conceived of individuals as robotic and emotionless actors (Lupton, 1999a: 10). Institutional reliance on quantitative scientific methods of risk assessment has led to a lack of sensitivity toward the cultural referents used by lay publics in developing understandings of risk (Mythen et al., 2000: 6). Of course, public institutions within capitalist cultures cannot be insulated against market forces and the drive for profit. As the BSE case infers, institutional 'solutions' to risk are generated in specific economic, political and organisational contexts (see Tacke, 2001).

NEGOTIATING TRUST

Having provided a favourable review and application of the risk society thesis, it is now necessary to gather together the objections which have been directed towards Beck's construal of the institutional definition of risk. In the remaining sections of the chapter I wish to bring the rougher edges of the risk society account into focus. The discussion generated here will serve to prise open the rather crude rendition of the lay–expert relationship, chasing up a number of issues only superficially quarried in the risk society perspective. Firstly, we will turn to issues of empirical validation, noting Beck's dependence on a narrow range of examples. This familiar problem will subsequently be expressed in specific terms through examination of the evidence of patterns of public distrust in expert institutions. From here, we will go on to argue that the risk society thesis presents a blinkered account of institutional performance and reproduces a realist understanding of risk. Finally, we will deal with the snags which arise out of Beck's ultra-structural approach, centring on the problems of institutional reification and cultural simplification.

Corresponding with the specifications of the risk society thesis, the BSE crisis indicates that public distrust in expert systems is a phenomenon that is gathering momentum. Nonetheless, although the risk society synopsis tallies with current academic assumptions, broad-spectrum trends cannot be established on the basis of discrete – or indeed clustered – examples. In delineating patterns of public distrust, Beck is over-reliant on arresting incidents, such as BSE, Bhopal and Chernobyl. As much as these episodes serve well as symbolic moments in the evolution of the relationship between experts and the public, they cannot be passed off as robust indicators of widespread distrust in risk-regulating institutions. Admittedly, the BSE crisis illumines public disquiet with expert systems, but this does not mask the fact that the risk society thesis is unnecessarily reliant on anecdotal offerings and remains woefully short on empirical substantiation. Establishing public distrust in expert institutions requires much more than meta-theory and a smattering of evidence on the side. Regrettably, Beck clumsily converts public responses to particular situations into a general and pervasive process of distrust. Whilst it is expectable that incidents of institutional mismanagement will encourage public scepticism, as we shall see in Chapter 7, trust is a complex and culturally rooted phenomena. With this caveat in mind, it is important to reflect on both the evidence *for* and the content *of* institutional distrust.

So, to what extent is the risk society thesis borne out by empirical findings? Do western publics increasingly distrust the relations of definition? In support of Beck, Anderson (1997: 113) cites a 1995 MORI survey which indicated that well under half of the British population had either a 'fair' or a 'great deal' of trust in scientists working for industry or government. Since this time, a growing body of social science research has charted a marked decline in public trust in expert institutions (see Coote, 1998; Grove-White, 1998; Macnaghtan and Urry, 1998: 262; Prior et al., 2000: 111). This said, nascent patterns of public distrust cannot be transformed into blanket rejection of expert knowledge. On the contrary, the research findings are far from unequivocal, signalling the need for interpretative delicacy. Drawing on a UK environmental research study, Dickens (1996: 95) refutes the suggestion that public trust in science has been eroded. Referring to large-scale survey data, Dickens notes that over a third of respondents believed that science offered a 'good explanation' of the relationship between individuals and the environment, without any critical qualifications. The bulk of respondents believed that science offered a 'good explanation' with a variety of qualifications, such as 'it is often used by governments and industry' and 'there are conflicting views within science'. Strikingly, less than 3 per cent of the sample believed that people should 'reject or be very suspicious of science' (Dickens, 1996: 96). These findings question the depth of public mistrust in science and indicate that the relationship between lay publics and scientific experts is not as cut and dried as Beck implies. The fallibility of scientific rationality has become a well publicised topic of discussion, but we cannot infer from this that the scientific paradigm has been rejected lock, stock and barrel. In contrast, many people still see science as a constructive and authoritative source of information about risks (CSEC Report, 1997). To rework the metaphor, it would appear that the 'milk money' belonging to science has been mislaid, rather than irretrievably lost. Instead of appreciating a plurality of standpoints and presenting a range of evidence, Beck has a propensity to imbue the general public with his particular version of reality. As Dickens (1996: 100) sagely warns, 'it is tempting to suggest that the thoroughgoing critique of science in which many critical sociologists regularly engage is being inaccurately projected onto the population as a whole'. Thus, the relationship between the lay public and scientific agencies is more complex than the risk society story would have us believe. In seeking to accentuate the difference between designated epochs Beck gives the impression that public disquiet about unbounded techno-scientific development is a novel occurrence. This is misleading, given that the rocky relationship between science and society has evolved over at least four centuries (Cohen, 2000: 154). As Goldblatt (1995: 177) notes, 'anti-technological social movements have a longer history in the West than the last few decades. A suspicion of science and its equation with progress is a deeply rooted western tradition that is as old as the original enlightenment celebration of science.' Furthermore, attitudes to science are historically and culturally embedded and relate to distinct incidents and national policies. Drawing upon comparative European research, Cohen (2000: 156) found that Italian respondents were more likely to see science as a benevolent endeavour than Germans, whilst the British rated science more highly as a tool of social advancement than the Danes or the Dutch. As far as the risk society thesis is concerned, we should 'tread carefully when trying to generalise about the relationship between science and society from what are German inspired theoretical insights' (Cohen, 2000: 173). Rather than routinely rejecting expert claims, lay publics may engage with scientific information in diverse and sometimes contradictory ways (Dickens, 1996: 101; Tansey and O'Riordan, 1999).[4]

Against the totalising approach of the risk society thesis, the more productive question may not be whether the public 'accept' or 'reject' science per se, but which underlying factors, qualifications and influences inform public understandings of science. On a wider note, public trust in social institutions should not be conceived as an 'either-or' choice. Even in seemingly clear-cut cases such as the BSE crisis, a variety of responses to institutional performance emerged amongst the British public (see Reilly, 1999). It follows that, instead of conceiving of public trust as an absolute which institutions either *possess* or are dispossessed *of*, we are better served by approaching trust as a process of negotiation. Expressed in this way, it becomes clear that we need to unpack the different dimensions of public trust, before speculating about the extent to which it is currently (dis)invested in expert institutions. Unfortunately, Beck fails to grasp that trust means different things, to different people, in different places, at different times. Far from retaining universal meaning, trust is comprised of an aggregation of factors, including delegation of responsibility, investment of faith, informational credibility, account-ability and confidence to perform. To blithely state that the public either 'trust' or 'distrust' social institutions succeeds only in collapsing the assorted dimensions of trust and silencing diverse attitudes and perspectives. Thus, we need to remain receptive to the ambiguities and complexities arising out of the dynamic interactions between the public and expert systems. Although further inquiry is needed to establish both the anchors and the dispersal of patterns of public trust, it is probable that outright distrust in risk-regulating institutions is more exceptional than Beck concedes.[5] State institutions may have been opened up to more intense forms of public criticism in the West, but this does not amount to wholesale rejection of their essential structure and/or functions. As Taylor-Gooby's (1999: 192) study of public attitudes towards social welfare reports, 'there is a pragmatic acceptance that people will be required to take more responsibility for meeting their needs, but at the same time there is a strong aspiration for government to play a stronger role'.

The realist tilt of the risk society argument flagged in Chapter 2 also resurfaces in relation to the social construction of risk. As one would expect, Beck is keen to disinter the economic factors that underpin the dissemination of information by expert systems. Of course, the balance between disclosure of risks and the protection of economic interests is a precarious one for capitalist institutions to negotiate. In an increasingly competitive world, social and welfare concerns regularly end up playing second fiddle to the economic agenda. Beck's claim that lay publics receive the most economically and administratively convenient methods of risk management, as opposed to the most effective ones is not without vindication (see Thompson, 1989). Nonetheless, Beck overstates the case by turning a blind eye to the capacity of institutions to provide culturally relevant and useful information about risk (Dean, 1999: 144). In the risk society narrative, scarcely concealed economic priorities set institutions out of synch with public demands: 'technology and natural science have become one economic enterprise on a large industrial scale without truth and enlightenment' (Beck, 1995a: 119). In such passages, it appears that the ultimate purpose of 'scientific rationality' is to mislead the public with the intention of ensuring ripe conditions for profit. While the connections between science and big business are well established, at times Beck comes close to dogma. It must be remembered that scientific investigation has produced a number of social benefits. The development of vaccines for tuberculosis, whooping cough and meningitis has led to the reduction of risk to public health. Furthermore, it cannot be reasonably asserted that 'scientific rationality' is deliberately deceptive, while 'social rationality' is collectively invested with truth. In actuality, both 'scientific rationality' and 'social rationality' are fallible discourses. As the Brent Spar conflict illustrates, where risk is concerned, 'truth' is a greasy commodity.

As a way of extending the explanatory potential of the risk society thesis, it is necessary to dial the contingent and organic nature of knowledge in to the equation. In preference to dealing in absolutes, it is more profitable to see truth about risk located along a continuum. To indulge in a metaphor, we might think of information about risk being deposited at various points of a piece of rope. The piece of rope is the subject of a tug-of-war, contested by experts from within and outside the relations of definition. To further confuse matters, the public may tug both *for* and *against* the relations of definition. At certain points one 'team' may possess more than the other. However, the rope is not static, being prone to violent tugs in the opposite direction. Moreover, in this game, team members are not averse to swapping sides, leading to sharp swings in perceptions and expectations. Metaphors aside, the point is that knowledge about risk can never be certain or final. In opposition to Beck, we should not assume that experts within the relations of definition are the bearers of misinformation about risks, any more than pressure groups and academics are the carriers of objective truth.

THE RELATIONS OF DEFINITION: DEHUMANISING THE PRODUCTION OF RISK

One of the key problems with the risk society narrative is that it approaches risk production in an transcendental fashion. As has been noted, the risk society thesis infers that societal risks are both manufactured and distributed by the relations of definition (Beck, 1995a: 110). Of course, if we latch on to certain risks such as nuclear power or iatrogenic illness, this is certainly true. Social institutions *do* produce risks, and, historically speaking, there is a growing recognition of this at the level of public knowledge. However, whilst the relations of definition are often implicated in the generation of dangers, it would be erroneous to perceive social institutions as isolated factories of risk production. Even a cursory glance over Beck's favourite vignettes of risk – air pollution, BSE and genetic technology – gives away that individuals and groups acting outside the relations of definition have been important players in both the production and social construction of risk.

Pace Beck, risk situations are more diverse, complex and multi-dimensional than the risk society narrative implies. Although the concept of organised irresponsibility provides a fitting insight into the defensive strategies of institutions, it encourages Beck to amplify the institutional production of risk and fade out human culpability. This shortcoming is exacerbated by the tendency to approach the relations of definition as an institutional block, rather than a field of interactive relationships. For instance, in *Ecological Politics in an Age of Risk* (1995a), Beck claims that hazard situations arise 'from the connection between economy and science, economy and law, economy and state' (Beck, 1995a: 182). Such a take on institutional power conjures up parallels with Weber's (1930) 'iron cage' of bureaucracy. By heavily inflecting the structural functions of the relations of definition, the risk society thesis veers decidedly close to reification. To conceive of the relations of definition as an anonymous power block eviscerates cultural institutions and fails to acknowledge the creative day-to-day activities of social actors. To infer that institutional structures act in issues of risk – rather than a *conglomeration* of individuals, positions and processes – glosses over the indubitable truth that public institutions are peopled by animate, cognitive individuals. Beck does make occasional reference to the active role of individuals within the relations of definition, but the overarching drivers of action remain hierarchies and occupational roles rather than volitional human beings (see Beck, 1992: 155–83).

As will be explicated in Chapter 5, the reification present in the risk society narrative is exacerbated by an unrefined conception of the lay–expert relationship. In particular, Beck's undifferentiated approach towards risk perception assumes a uniformity of cultural experience which is not supported by empirical evidence. As Wynne (1996) argues, Beck overplays the divide between the good citizens of the lay public and malevolent experts working within social institutions. This deficiency is typified by the rather crude separation between 'social' and 'scientific' rationality (Beck, 1992: 30). Beck imagines social dysfunction to be a direct product of the development and application of scientific rationality. Yet such a conspiratorial attitude detaches science from any form of meaningful relationship with the public. To some extent, scientific and technological discourses and practices have evolved in relation to the needs, demands and aspirations of citizens. As we shall see in Chapter 7, the boundary between expert and lay knowledge is more fluid and dynamic than Beck is willing – or able – to acknowledge. It must be recognised that ideas and values about risk are publicly generated as well as institutionally disseminated, with lay and expert groups interfacing, rather than acting as fixed and frozen boundaries (Wynne, 1996: 76). Alas, Beck fails to recognise that the 'done to' lay public are at one and the same time 'doers' working within the relations of definition.

The rigid lay–expert demarcation contributes towards a rather caricatured depiction of the relationship between structure and agency in the process of risk definition. As will be made apparent in Chapter 8, high levels of bureaucracy within institutions can serve to police and restrict what is 'sayable' and 'doable' about risk. Nonetheless, at each stage of the chain of risk – in manufacture, definition and regulation – human beings are present and, to varying degrees, active in the decision-making process. In assigning overarching power to the structure of the relations of definition, Beck abrogates responsibility for social risks from individuals, save an elite band of vaguely defined 'experts' acting within the confines of predetermined institutional guidelines. It is perhaps worth reminding ourselves that scientific and economic methods of hazard calculation – such as the calculus of risk – did not arbitrarily appear in western culture. Social institutions employing economic and scientific principles have, at least notionally, been built with public blessing. It is probable that a significant proportion of individuals in the West are generally supportive of the goals of public institutions, particularly when set against the backcloth of less egalitarian regimes around the globe. Without misguidedly celebrating the merits of a divisive and exploitative system, we should recognise that capitalism does provide for the basic needs of the majority in western cultures.

At a minimum, we must acknowledge that institutional methods of managing risk have developed recursively, as a result of social conflict and social consent. For certain, some parties are more culpable than others in relation to the ideological goals of social institutions and the mismanagement of risk. However – taken as an ahistoric collective – the general public have been complicit in the development of methods and mechanisms of risk management. Ultimately, only *people* can people institutions. Thus, more than a grain of responsibility for the current crisis of risk management falls in the lap of us, the public. As discussed in Chapter 2, it is not only institutions which produce environmental risks, but individuals themselves. Beck fails to grasp that – in the production and regulation of risks – the public can be cast as both victim and accomplice.

A final flaw in the risk society argument emerges around the issue of cultural variability. In methodological terms, Beck's amorphous depiction of the relations of definition cannot possibly do justice to the changeability of social institutions in different nations. Beck's borrowed notion of the safety state may travel some way in relating the development of German – and, at a push, British – welfare systems. However, it is not the story told in other regions, such as Southern Europe, Scandinavia or North America.[6] The extent to which technological, scientific and cultural practices are deemed to be risky will be subject to regional variations. Not only are normative attitudes towards risk variable, legal policies are also marked by cultural difference.[7] The risk society thesis may speak to the experiences of certain groups living in mature social democracies, yet it remains a non-generalisable argument (Marshall, 1999: 267).

CONCLUSION

In the course of the chapter we have charted the role of dominant institutions in the identification, construction and regulation of risk.

In drawing this discussion to a close, it is clear that Beck's conception of the process of risk definition has both benefits and weaknesses. Indisputably, the risk society thesis offers a compelling insight into the relationship between dominant social institutions and the production of risk. Beck is justified in pointing out that certain types of risk are beyond the legislative powers of the nation state. Manufactured risks and uncertainties are often unearthed post hoc, escaping scientific, legal and medical regulation. As the BSE case demonstrates, the inability of the relations of definition to cope with manufactured risk is wont to invoke various forms of institutional obfuscation. In particular, the concept of 'organised irresponsibility' enables us understand how and why responsible institutions have directed public attention towards image management and away from the details of preventative measures. Technically, the concept of organised irresponsibility retains utility as a tool for dissecting institutional approaches to risk. Theoretically, organised irresponsibility captures the incommensurability between existing systems of risk management and the production of complex risks.

Nonetheless, the deployment of the relations of definition as a mechanism for understanding the social construction of risk is not without problems. First, Beck's argument falls back on eye-catching examples of institutional failure and is empirically disengaged. Second, in application, the idea of relations of definition reifies and depopulates social institutions. Third, the self-imposed rigidity of the relations of definition steers the risk society narrative towards an uncultured separation between lay and expert groupings. Fourth, the idea of relations of definition is too blunt a tool of analysis to capture cultural variations between institutions in western nation states. Fifth, the risk society argument can be criticised for talking down the social benefits provided by institutions. What is absent it seems is due account of the context in which risks are defined and mediated. Public scepticism towards expert systems cannot be causally attributed to institutional failure to regulate risk. Set within a milieu characterised by smooth interchange of people and products and unfurling global communication networks, we should not be surprised that more extreme forms of public scrutiny are in the ascendant. All the same, the intensification of public criticism should not be mistaken for wholesale rejection of state institutions (Boyne, 2003: 88; Taylor-Gooby, 1999: 193). The combined weight of these misgivings leaves the risk society thesis looking rather bruised, urging a more nuanced account of the relationship between the public, government and science. The abstract and sweeping nature of Beck's argument tramples indelicately over the idiosyncrasies and intricacies of lived cultural experiences. As we shall see, nowhere is the disparity between theory and everyday practices more apparent than in Beck's account of the role of the media in communicating risk.

4
Mediating Risk

Up to press, we have discussed the generation of risk in contemporary western society, both as a material product and a conceptual category. Sketching out a silhouette of the risk society thesis, in Chapter 2 we traced the physical production of risk, noting the contested nature of environmental effects. In Chapter 3, we turned to the institutional construction of risk, recounting the dominant role of science and government in the processes of identification and regulation. With reference to the collective relations of definition, we explained how and why risk-regulating institutions are currently struggling to cope with the velocity of manufactured risks. It was also noted that Beck's reading of the definitional process foregrounds the operations of government and science as primary shapers of risk meanings. Here we will argue that – despite their being key sites of risk definition – Beck's over absorption with science and the state leads to theoretical marginalisation of other important agents of risk communication, such as the mass media.[1] The prime sequale of this chapter is to demonstrate that the media is a crucial mechanism of risk communication in contemporary culture and one which is undervalued in the risk society thesis. In the following pages, we will seek to explicitly redress the balance by considering the functions of the mass media in the process of risk representation, concentrating on the contextual features which underpin media operations. In the opening section, we expound Beck's controversial casting of the media in the risk communications process. In the second section, we begin to pick out the holes in the risk society account, demonstrating that patterns of ownership and control and economic power vitally influence media output. From here, we go on to assess the extent to which internal production processes filter what can be 'said' about risk, pulling out the ramifications of media selectivity for the provision of public information. Finally, forming a bridge to Chapter 5, we begin to unravel the link between media representations and public understandings of risk, questioning the capacity of the media to perform as a lever for oppositional action.

THE MEDIA IN THE RISK SOCIETY

Over the last 50 years, the mass media has been recognised as a primary source of public information about risk (Anderson, 1997; Friedman et al., 1996: 18; Stallings, 1990). In contemporary western cultures, media portals act as important sites of knowledge, advice and debate. Widely mediated crises such as 9/11, the invasion of Afghanistan and the US-led attacks on Iraq have transported risk into our living rooms in 'real time' with jaw-dropping clarity. In a media-saturated culture, the communication and representation of risk have materialised as a hot topic of discussion.

Sidelining the issue of risk for a moment, it is apparent that contemporary social discourses are informed by media symbols, images and meanings (Beck, 1995a: 9; Reilly, 1999: 188; Stevenson, 1999). Academic acknowledgement of the media as an increasingly pertinent source of information and sense making has led to an upsurge of interest in media functions and effects. The changing shape of electronic, digital and satellite technologies is reforming the relationship between the media and the public. Recent trends towards software synergy, coupled with a drive toward hardware convergence have further transformed the landscape of information and communication flows. As the cultural profile of media expands, those able to buy in to the so-called 'technological revolution' are able to navigate a veritable vortex of communication systems (du Gay et al., 1997; McNair, 1998). Nevertheless, amidst the fanfare which has regaled technological advances, underlying power relations have been augmented rather than surmounted. The ownership and control of the mass media have contracted, the gap between the 'information-rich' and the 'information-poor' has widened and the western media has extended its global reach. This turbulent mixture of technological change and reinforced control provides the context in which the mediation of risk in contemporary society should be understood. At the same time as new media technologies have potentially diversified routes for the dissemination of risk communications, traditional media such as mass circulation tabloids and television news programmes have retained their status as crucial sites of public risk information.

Since the late 1980s, traditional media outlets have increasingly hooked onto risk as a topic of concern, leading to a general intensification in the coverage of risk-related affairs. In modern western society, the mass media is an indispensable machine of risk identification and an important conveyor of strategies of safety (Anderson, 1997; Philo, 1999; Sjöberg and Wahlberg, 1997). The rising cultural profile of risk has led to the media performing an increasingly influential function in the processes of risk communication (see Cottle, 1998; Kasperson and Kasperson, 1996; Nelkin, 1987). At a general level, the upsurge of interest in risk within the news media can be validated with reference to quantitative indicators. Lupton (1999a), for example, performed a quick-and-dirty comparison of the use of the word 'risk' in national newspapers in Australia. In 1992 'risk' appeared 2,356 times in the main text and 89 times in headlines. By 1997, the term appeared almost 3,500 times in the text and in 118 headlines (Lupton, 1999a: 10). At one level, such observations are no more than cursory pointers which need to be framed within the context of an increasingly risk-aware, health-conscious and litigious social environment. At another, there can be little doubt that risk has become an area of heightened interest, both within the media and culture more broadly.

Following the hype surrounding the dispersal and effects of the mass media, we should not be unduly stirred by Beck's summation that, 'the risk society can be grasped theoretically, empirically and politically only if one starts from the premise that it is always also a knowledge, media and information society at the same time' (Beck, 2000d: xiv). Taking the above quote on board, we might logically expect the media to be absolutely central to the formulation of the risk society hypothesis.[2] However, curiously, media operations are largely conspicuous by their absence. As I shall go on to demonstrate, despite intermittently ascribing great informational power to modern communications systems, Beck under theorises media structure and overloads media effects.

In *Risk Society* (1992: 197) mediated communication is described as fundamental to the formulation of public understandings of risk. Beck's endorsement of the mass media as a decisive channel of risk information has been restated and reinforced in recent years (see Beck, 1999; 2000e). Nevertheless, Beck's investigation of the media remains unquestionably scattered, rather than systematic (see Anderson, 1997: 188; Cottle, 1998). Perhaps the fullest accounts of the role of the media in disseminating risk information can be found in *Ecological Politics in an Age of Risk* (1995a) and *The Anthropological Shock: Chernobyl and the Contours of the Risk Society* (1987). In both of these works, Beck emphasises the responsibility of the media to inform the public about undetectable dangers and hazards. It is argued that, in the risk society, the invisible quality of manufactured risks induces a form of 'cultural blindness', which works 'downright mysteriously, since nothing has changed for the eyes, nose, mouth and hands' (Beck, 1987: 154). Due to the intangible quality of manufactured risks, people around the world may unknowingly and unwittingly suffer from the deleterious side effects of manufactured risks, such as air pollution or food contamination. Because the health effects of such risks are both delayed and disguised, the mass media becomes an increasingly prominent font of information for the public.

Despite episodically mentioning the transformatory potential of new media technologies, Beck allots the media an ambiguous role in the social construction of risk. In the risk society perspective, prevailing institutions rely upon the media to transport risk information. Yet the media also retains a degree of autonomy, being somewhat detached from the operations of the relations of definition. Hence, the media has the capability to function both as a vessel for institutional information and as a mouthpiece of public critique (Beck, 1992: 197) Paradoxically, the media is part of the relations of definition *and* the apparatus by which prevailing institutional power relations may be challenged. Oddly, Beck neither addresses nor justifies this bipolar approach, manoeuvring between the two positions according to the dominant trajectory of the argument. In certain instances, the media is portrayed as a vehicle for the translation of risk information from expert bodies to the lay public (Beck, 1995a: 96). In this guise, the media is very much part of the established relations of definition, channelling risk knowledge from scientific and governmental experts to the lay public. Performing from within the relations of definition, the media visualises the preferred messages of politicians, business analysts and elite scientists (Beck, 1992: 32). In accord with this position, the capacity of the mass media to uphold institutional values has long been acknowledged within sociological theory (Lodziak, 1986; Marcuse, 1964). In the 1970s, Cohen's (1972) celebrated moral panic model employed a 'deviancy amplification spiral' to illuminate the linkages between institutional stigmatisa-tion, media amplification and public perceptions of risk. A decade later, Hall et al. (1982) applied the moral panic model to the problem of street crime. Hall's much-cited study aptly demonstrates how the amplification of risk by politicians and the mass media was used to produce public acquiescence in governmental policy in Britain. At the time, a general climate of fear about rising crime rates enabled the Conservative government to push through fundamentally repressive law-and-order measures.

In reverse, Beck also wishes to maintain that the media have the capacity to perform outside the relations of definition, as a social siren and an agent of institutional opposition (Alexander and Smith, 1996: 255). In *Ecological Politics in an Age of Risk* (1995a) Beck explores the relationship between science, media discourse and public understandings of risk. As we shall see, in the risk society thesis, the mass media is depicted as both an *example of* and a *motor for* lay reflexivity. Media potential for institutional reflexivity is exemplified by counter-expert scientific and political information carried within the pages of newspapers and broadcast on news bulletins. In the risk society narrative, media products have the capacity to stimulate enhanced forms of public reflexivity. For Beck, the mass media has the power to destabilise and unbind the assumptions of scientific inquiry, challenging the legitimacy of governing institutions and encouraging public distrust in expert systems (Beck, 1992: 154). In this respect, publication of hazards by the mass media acts as a palliative to the 'cultural blinding' generated by the intangible quality of manufactured risks. By way of illustration, Beck refers to the public disquiet about nuclear technology provoked by the Chernobyl disaster: 'what would have happened if the mass media had remained silent, if the experts had not quarrelled with one another? No one would have noticed a thing' (Beck, 1987: 154). Beck believes that media coverage of Chernobyl enabled counter-hegemonic voices to be articulated, raising suppressed questions about the safety of nuclear and chemical technologies. Manufactured risks such as Chernobyl are described as a 'pure media events', which serve to orchestrate public dialogue (Beck, 1995b: 96). Following this plot, the media is sanctioned as the discursive space in which political contestation about risks takes place (Beck, 1992: 46). Cast as an oppositional player, the media operates as a 'public watchdog', guarding against institutional corruption and championing human rights. In such a scenario, the media possesses the power to challenge the dominant relations of definition in the production, identification and management of manufactured risks (Beck, 1995a: 140). This liberal pluralist position is counterbalanced by occasional reference to the economic context in which media outlets operate in the West. For instance, in *Risk Society* (1992: 126), Beck notes that the media are, 'limited and checked by the material conditions on the production of information and the general legal and social conditions'. There can be no doubt that such a mixed depiction of the media reproduces inconsistencies within the argument. Beck tends to oscillate between two diametrical poles, leaving it unclear whether the media is best understood as an emancipatory vehicle of the people or as a tool of state propaganda (Cottle, 1998: 9).

This said, it would be misleading to infer that the weight of emphasis between the two poles is evenly balanced. Beck's political optimism unquestionably inclines him toward the oppositional potentialities of the mass media. The emancipatory trajectory of the risk society thesis dictates that the media is dominantly cast as a public guardian and consciousness raiser. As Alexander and Smith note: 'The result of increased media focus as Beck sees it, would be the increase in objective information, and he appears confident that this information will automatically register on contemporary consciousness' (Alexander and Smith, 1996: 255).

For Beck, the dissolution of nation-state control, increased global information exchange and the diversification of new technologies have opened the media up to the opinion of protest groups, counter-experts, maverick scientists and lay public testimonies (Beck, 1995a: 141). Media representation of public opinion – or 'social rationality' – serves to undermine scientific and economic logics of risk promulgated by government and big business. Because contemporary risks are oblique, the mediated voices of counter-experts are indispensable in facilitating democratic speech. By pursuing and criticising institutional perspectives on risk, the media serve to 'explode hazards', making dangers distinguishable to the public.

Although the dualistic quality of the risk society narrative allows Beck room for manoeuvre, his core assumptions about the capacity of the media to catalyse public reflexivity are difficult to substantiate. Undoubtedly, media outlets do have the ability to select and frame risk issues and to act as vehicles for public discussion.[3] However, Beck's exaggerated style sporadically propels him toward media-centrism. At times, it seems that everything turns on the mediation of risk. For instance, in 'The Anthropological Shock: Chernobyl and the Contours of the Risk Society' (1987), Beck asserts that the transference into the risk society entails the 'end of perceptiveness' and the 'beginning of the social construction of risk realities', where 'information equals reality' (Beck, 1987: 156). Of course, Beck's practical desire to generate debate about changing methods of risk communication is to be applauded. Nevertheless, such sweeping claims disrupt the tempo of the risk society argument. If the media holds great sway in the social construction of risk, why is it afforded such infrequent attention by Beck? Equally, just how much does Beck's work tell us about the critical process of meaning making? At a theoretical level, the risk society thesis lacks sensitivity to both the socio-economic construction of risk and the manner in which media information is interpreted by autonomous individuals. Indeed, Beck remains some distance from important debates which have preoccupied media and cultural theory in recent years, including those around media production, ownership and control and cultivated audiences. As these exchanges illustrate, the ability of the media to uniformly represent and communicate risk cannot be taken for granted. Unfortunately, Beck tends to look right over the economic structure of media industries and abstracts media outlets from their working cultural contexts. In order to advance a fuller understanding of the social construction of risk, we need to arrive at a more sophisticated account of the structure and functions of the mass media in contemporary society.

MEDIA CONTROL AND THE REPRESENTATION OF RISK

Having recounted the scarce but exaggerated presence of the media in the risk society, I now wish to expose the empirical gaps and theoretical weakness which destabilise the core argument. By his own admission, Ulrich Beck is a sociologist with no great expertise in media theory (Beck, 2000e: viii). This confession enables us to understand the indefinite and patchy positioning of the media in the risk society thesis. However, Beck's lack of familiarity with the subject does not mean that his assertions about the role of the media in contemporary society should be exempt from scrutiny. In the following sections, we will demonstrate that the risk society argument is blind to the political economy of the mass media and unappreciative of the everyday practices of journalists and reporters. As we shall see – given that primary forces such as the ownership and control of the media and the media production process act as vital filters to risk information – these are serious shortcomings.

So, how accurately does the risk society thesis reflect the activities of the media in contemporary culture? Specifically which economic factors impinge upon the process of risk communications? In responding to these questions, we will draw upon the theoretical contributions of Anderson (1997), Cottle (1998) and Hargreaves (2000). The empirical dimensions will be addressed through a huddle of case studies which have probed the relationship between media representations and public understandings of risk (Eldridge: 1999; Reilly, 1999; Reilly and Kitzinger, 1997).

In the risk society thesis, the mass media is periodically cast as a vital source of definition, contestation and information. By and large, the media is seen to act in the public interest, unmasking risks and challenging the relations of definition (Alexander and Smith, 1996: 255; Beck, 1992: 115). In as much as Beck is correct in observing that the media has the ability to question institutional evaluations of risk, analysis of the underlying economic factors that determine and shape representations of risk is patently absent. By upholding the crusading role of the media in heightening public consciousness, Beck presents an undeveloped and partial account. Lurking beneath this utopian vision rests the erroneous assumption that all occurrences have an equal chance of being reported. In the risk society narrative, the mass media is constantly lingering behind the arras, waiting to seize upon breaking risks and interrogate institutional procedures. However alluring in principle, this notion of the media as public watchdog fails to engage with a dense set of structural interactions which take place prior to mediated representations of risk. Exactly *which* risks become the focus of public concern and which escape scrutiny is critically dependent upon internal power relations and the flow of information entering and exiting media outlets (Bennett, 1998; Hargreaves, 2000). As Douglas (1985: 60) reasons, 'something is happening to fasten attention on particular risks and to screen out perception of others'. But what exactly is this 'something'? Why is it that seemingly minor risk events may receive disproportionate coverage in the media, whilst major catastrophes go unreported?

Taking a contextual step backwards, it first needs to be recognised that the production and distribution of news takes place in large hierarchical organisations that are technically complex and geared towards the generation of profit (Green, 2003: 221; Negrine, 1994: 118).[4] In western capitalist society, a myriad of media interests are owned and controlled by just a handful of individuals (Croteau and Hoynes, 2000: 38; Stevenson, 1999: 112). Thus, a formidable degree of economic and cultural power is possessed by proprietors of media empires, such as Ted Turner and Rupert Murdoch. The global companies owned by these media magnates are oriented towards 'synergy', or the integration of interlinked media industries and technologies (Negus, 1997: 84). In the future, it is likely that media convergence and the digitalisation of information will increase the geographical scope and economic influence of the elite few that own and control the global media.

Unfortunately, appreciation of the organisational and the economic context of media production does not make up part of the risk society tapestry. Beck is noticeably reluctant to acknowledge that news organisations are branches of vast global media organisations. This is an important omission, given that the concentration of media ownership has traditionally raised concerns about cultural and political domination (Croteau and Hoynes, 2000: 48; Tomlinson, 1997: 126). In a profit-driven media environment, Silvio Berlusconi – the current EU president and Italian prime minister – has demonstrated that he who owns most is invariably he who speaks loudest (see Boyne, 2003: 28). As a direct result of extra-media ownership, conflicts of risk interest have periodically surfaced. For example, during the Gulf War in 1991, General Electric owned a significant chunk of NBC News, whilst being simultaneously involved in the production of bomb parts used as weapons of destruction against Iraq. The implications of this state of affairs for objective news reporting of the Gulf conflict hardly needs spelling out. In another well documented case, Tiny Rowlands, the former owner of the *Observer*, outlawed reporting of civil unrest in Zimbabwe in an attempt to protect his economic investment in the country (Curran and Seaton, 1989: 93).

The continued privatisation of the media, combined with a rise in cross and extra-media ownership, restricts the ability of the media to function as a public watchdog. Many burgeoning media technologies, such as the internet and digital television, rely heavily upon advertising revenue to produce profit. Even the more traditional news media – such as national newspapers – are heavily reliant upon advertising revenue, which makes up approximately three-quarters of total profits for broadsheets, and just under half for tabloid newspapers (Barwise and Gordon, 1998: 20; Negrine, 1994: 67). In a delicate economic environment, conflicts inevitably arise between newspaper editors and advertisers. Obviously, it is not in the interests of large media organisations to discourage the sizeable revenues offered by advertisers (Collins, 1992; Green, 2003: 221). Naturally, this economic bind has important ramifications for the reporting of a range of issues, including risk. In Britain, state institutions are amongst the largest newspaper advertisers. Further, the British government grants the BBC its public service warrant and determines the level of the licence fee. These observations suggest that media representations of risk cannot be sequestered from the political and economic context of media production. As Negrine reminds us:

> The economic and political needs of media organizations – the need to survive, to maximize profit, to increase sales, to increase advertising revenue, to maintain a political line, to placate politicians – form an important backdrop to the study of the production of all media content. (Negrine, 1994: 118)

Regrettably, the cultural, economic and political context in which the media operates is not in attendance in the risk society thesis. As a consequence, a thorough understanding of the political economy of the mass media goes awry. In this respect, Beck's understanding of the media is at best embryonic, at worst, rather naive.

FILTERING RISK: THE MEDIA PRODUCTION PROCESS

Thus far, it has been argued that the risk society's free-floating depiction of the media fails to recognise that news media outlets are subject to a variety of economic and political forces which influence the reporting of risk incidents. In this section, we supplement our critique by considering the way in which the internal production process structures the nature and quality of mediated risk information. Accordingly, discussion will be oriented around editorial objectives, news values, selectivity and sourcing. Highlighting further elisions within the risk society thesis, it will be demonstrated that the media production practices and associated regulatory procedures shape the quality and range of publicly available information about risks.

On a daily basis, news organisations are dependent upon a constant stream of information (Palmer, 1998). One of ways in which news organisations ensure a ready supply of news is by routinising news flows. The routinisation of news requires – amongst other things – an accumulated bank of reliable and consistent sources (Schlesinger, 1990). In addition to political and economic factors, the reporting of risk by news organisations will be influenced by source availability and journalistic selection (Coleman, 1995: 68; Miller and Riechert, 2000: 45; Reilly and Kitzinger, 1997: 324; Schlesinger, 1990). As a means of accumulating news, journalists are routinely placed within information-bearing institutions such as law courts, police stations and parliament.[5] In Britain, selected news reporters sit in the lobby area of the House of Commons where they are free to mingle with government and opposition ministers. For many political journalists, the lobby is a major source of information about risk. Yet alliances forged in the lobby can be fragile and volatile. Every so often, politicians mislead reporters and journalists misquote sources (Negrine, 1994: 134). Of course, those granted lobby privileges are understandably wary of offending potential news sources and this alone raises the issue of professional objectivity. The distance between journalists and politicians has contracted still further in recent years with the emergence of 'spin-doctors' and 'issues management' consultants who ply their trade in the interstice between media and politics (Sparks, 2003: 199). Due to the reciprocal relationship between reporters and politicians, the scope for oppositional reporting of risk is limited. Journalists that adopt anti-governmental positions are likely to find their supply of information cut. In a competitive profession, the possibility of 'losing news' will encourage reporters to choose their enemies carefully. This well rehearsed game of cat and mouse has significant consequences for what materialises as news, demonstrating that the quality of public information about risks is affected by idiosyncratic relationships between politicians and media professionals.

Outside the etiquette of the lobby system, research indicates that institutional sources are key agents in the formation of media representations of risk (Miller and Riechert, 2000: 45). As one might expect, power brokers within science and government have been found to be the most frequently used sources by journalists reporting on risk issues (Coleman, 1995: 68). However, as Kitzinger and Reilly (1997: 319) warn, reliance on press releases distributed by agencies involved in risk management can encourage partial news reporting.[6]

In addition to garnering information from institutional sources, news organisations also routinise news flows by reporting on set activities from the 'diary'. Again, the dependence of news journalists on diary events impacts upon the reporting of risk. As Anderson (1997: 120) notes: 'some news stories have a much greater likelihood of being covered than others because they accord with organisational norms, pressures and routines and/or they possess particular conventional features'. Far from arbitrarily reporting risk events, journalists will turn to reliable sources – such as scientists and politicians – who have provided information in the past. The constant pressure of deadlines steers journalists towards building up a small number of well known contacts from within the relations of definition, rather than drawing from an amorphous range of sources (Anderson,

1997: 129). This dependence on trusted institutional commentators is borne out by empirical studies of media content. Focussing on news coverage of environmental risks, Hansen (1990; 1991; 2000) reports a clear imbalance in the ideological range of sources utilised by journalists. Using comparative analysis of environmental affairs in Britain and Denmark, Hansen (1990) discovered that 23 per cent of primary sources were drawn from public authorities, 21 per cent from government, 17 per cent were attributed to independent scientists and just 6 per cent were representatives of environmental organisations. Hansen's (2000) more recent work confirms that news reporters continue to be unhealthily dependent upon institutional sources in constructing reports about risk.

The evidence harvested from various strands of research indicates that Beck's anticipated 'media explosion' of risks is far from guaranteed. Media outlets are quite capable of opposing the status quo, but in practice this tends to happen only up to a critical point.[7] It must be appreciated that a significant chunk of news reporting relies upon the input of institutional informants, with many journalists regularly reporting from within the relations of definition. Ipso facto, boundaries are occasionally stretched, but rarely transgressed. As a consequence of the routine practices of journalists within media organisations the interests of dominant groups are not habitually opposed.

A further salient feature of the media production process missed by the risk society thesis is the reproduction of 'news values'. Stuart Hall (1973) famously described news values as a set of assumptions based upon knowledge about the audience, dominant assumptions about society and a professional code or ideology. In order to maintain audience interest and cultural relevance, news reports must broadly fit the criteria of 'newsworthiness'. The contents of newsworthiness will, of course, vary over time and place. In a seminal study, Galtung and Ruge (1974) identified twelve decisive news values, including frequency, amplitude, cultural relevance and degree of personalisation. Events which possess the greatest number of news values have the highest probability of being reported (Palmer, 1998: 378). Despite changes in the characteristics of news, elements of Galtung and Ruge's theory have retained significance. With specific reference to risk, Greenberg et al.'s (1989) content analysis of American television suggests that media coverage of risk incidents tends to follow the pattern established by news values, being directed by 'events' rather than 'issues'.[8] By concentrating on images and presenting risks as events journalists are able to 'frame' stories and to side-step sticky contextual issues (Anderson, 1997: 21). Logically, news journalists will gravitate towards spectacular and emotive incidents which can be readily visualised (Boyne, 2003: 107; Sjöberg and Wahlberg, 1997: 4). As Allan et al. (2000: 9) appraise, news coverage is 'event-centred as opposed to issue sensitive, the main result being that those potential sources capable of placing the event in question into a larger context are regularly ignored, trivialised or marginalised'. Again, this denotes that news outlets will tend to be discerning in their coverage of risky incidents.

Reworking Galtung and Ruge's news values, Peter Bennett (1998) has developed a useful framework for predicting media reporting of risk. According to Bennett, the likelihood of a risk being reported can be indexed to the presence or absence of 'media triggers' (1998: 16). These triggers include evidence of cover-up, blame, human interest, conflict, signal value and visual impact. Although thorough application is beyond the ambit of this chapter, research has demonstrated that the magnitude of media coverage can be gauged by employing Bennett's triggers (see Mythen et al., 2000: 41–4). For both Bennett (1998) and Galtung and Ruge (1974), the degree of ambiguity is of central importance in determining news output. The less contradictory the information, the greater the likelihood that it will be translated into news (Negrine, 1994: 120). In covering complete and unequivocal events, newspaper journalists find themselves in the informational comfort zone. Ill-defined and ongoing issues, however, are less easily mediated. Journalists working under the constant pressure of deadlines may be tempted to construct 'certainty' about risks – even at the cost of accuracy: 'Although headline writers mostly deal with the appearance of certainty, they know that there is no such thing as a single indivisible "truth" ... but the communication of uncertainty doesn't sit easily with three-word, 72-point headlines' (Hargreaves, 2000: 3).

The lack of definitional clarity that surrounds breaking risk incidents means that news journalists are charged with the unenviable task of lucidly representing oblique and changeable situations (Friedman et al., 1999). Such trying criteria may lead to risk issues being marginalised, under reported or simply ignored. The harsh journalistic demands of accuracy and punctuality enable us to understand why the degree of media coverage cannot always be equated with the harmfulness of risk (Hansen, 1990; Kitzinger and Reilly, 1997: 320; Macintyre et al., 1998). With some justification, Barbara Adam (2000b) calls attention to the problems of media processing which arise out of reporting ongoing threats to the environment:

> Environmental 'news' therefore, is almost a contradiction in terms ... the long-term continuous pertinence of possible danger constitutes a major challenge to newswork: news as the delimited here and now of events has to be rethought in the context of the long-term and continuous manufacture of these hazards as inescapable by-products of the industrial way of life in general. (Adam, 2000b: 122)

Contra the risk society thesis, we need to distinguish between the coverage of *different* risks in media reporting (Cottle, 1998: 17). Far from being uniformly exposed, risks work along a sliding scale of cultural resonance (Bennett, 1998; Hansen, 1991). On top of this, a whole sequence of internal production practices intervene between risk incident and headline news:

> Every newspaper when it reaches the reader is the result of a whole series of selections as to what items shall be printed, in what position they shall be printed, how much space each shall occupy, what emphasis each shall have. There are no objective standards here. (Lippmann, 1965: 223)

In failing to probe the professional culture of media organisations, the risk society thesis also glosses over the backgrounds and morals of journalists and news reporters. Suffice it to say, not all journalists fit the description of leftfield, environmentally concerned citizen. As Negrine (1994: 129) points out, 'journalists, like everybody else, carry ideological baggage and so cannot report events in some pure or universally truthful way'. This is not to sweepingly declare that journalists and broadcasters are wanton apparatchiks of government, but neither should we assume that they are politically motivated oppositional agents. What is certain is that media professionals will be obliged to self-censor, according to editorial/proprietorial margins. Although Beck's risk society thesis depicts a media devoted to objectively uncovering environmental risks, event-centred journalists will be selective in representation, sometimes to the detriment of precision (Eldridge, 1999; Singer and Endreny, 1987). As a consequence, certain risk situations may be mediated in an exaggerated or distorted fashion (Anderson, 1997: 115; Laurance,

2000; Sjöberg and Wahlberg, 1997: 4). Having said this, it would be wrong to argue that journalists deliberately set out to mislead the public in the reporting of risk (Jones et al., 1997: 8). Allowing for the multiplicity of possible readings of risk events, ambiguity is a familiar occupational hazard. As Hargreaves (2000) explains, the indeterminate character of environmental risks makes them extremely difficult to pin down in compact news reports. This presents journalists – not necessarily schooled in environmental and scientific affairs – with the daunting problem of what to 'say' about risks (Wilson, 2000: 201). Predictably, empirical studies have demonstrated that a significant degree of confusion exists amongst journalists covering environmental issues. Bell (1994) notes that media reports often fail to distinguish between, or simply conflate ozone depletion and the greenhouse effect. In direct opposition to Beck's celebration of consciousness raising through media coverage of Chernobyl, content analysis of television newscasts about the disaster revealed that reporting was largely uninformative, contradictory, poorly contextualised and lacking in comparative examples and figures (see Friedman et al., 1987; Wilkins and Patterson, 1987).

It can be argued then that the risk society perspective fails to grasp the routine modes through which risk issues are presented in news reports. This is particularly apparent in relation to the construction and representation of expert and lay discourses (Cottle, 1998: 19). It has been taken as read within media theory that broadcast news tends to be imbued with an ideological slant supportive of dominant social groups (Hall, 1973; Negrine, 1994). Whilst this ideological bias may be less palpable since the introduction of narrowcasting, it still exists within many broadcasting networks. Consequently, risk issues are often presented in a formulaic manner in the media, particularly in television news bulletins. In many respects, the dominant formula of presentation in newscasts mirrors the narrative structure of an epic film drama. The report may begin with the manifestation of the risk as *problem*; say, the continued manufacture of genetically modified maize in Britain. It may continue to outline the *conflict* which has arisen following the initial identification of the risk; for example, Greenpeace protesters attempting to destroy genetically altered crops in test centres. We then move on to focus upon the doleful *victims* of the conflict; the farmers whose crops have been needlessly destroyed. Finally, the *resolution* – or 'happy ending' – arrives, as government experts dismiss campaigners as alarmist, refute any evidence of harm and promise greater protection of GM test sites in the 'public interest'.

Of course, this is not to suppose that lay publics passively accept preferred readings of television news. As we will argue, the ideological effects of news reporting can only be properly gauged by accessing the meanings made by audiences. Nevertheless, the prevailing structure of broadcast news reports does seem somewhat removed from imagining the media to be an oppositional agent. Despite Beck's insistence that media investigation reveals inaccurate governmental assessments of risk, the inverse also applies. Amidst public fears about the Aids epidemic in the 1980s, the British government provided valuable information and advice at the same time as sections of the tabloid media were engaged in a crusade of wilful misinformation. Making direct reference to the reporting of Aids, Eldridge notes that several tabloid newspapers initially denied that heterosexuals were at risk and were disparaging about safe-sex campaigns. A *Sun* newspaper leader entitled 'AIDS – The Facts, Not The Fiction' ran as follows:

> At last the truth can be told. The killer disease AIDS can only be caught by homosexuals, bisexuals, junkies or anyone who has received a tainted blood transfusion. *Forget* the television adverts, *Forget* the poster campaigns, *Forget* the endless boring TV documentaries and forget the idea that ordinary heterosexual people can contract AIDS. They can't.[9]

This editorial serves as a stark example of the media's potential to obscure issues of risk and to misinform the general public. It might be added that the publication in question is currently Britain's largest-selling daily newspaper. The reporting of risk in tabloid newspapers also picks out another blemish in Beck's understanding of the media. Insofar as the risk society thesis refers to the media as a homogeneous body, it is worth pointing out that information about risks is mediated through distinct communication channels (Cottle, 1998; Hargreaves, 2000). As van Loon (2000b: 234) asserts, 'the media environment consists of a multiplicity of forces that may not always pull in the same direction'. Representations, images and ideas about risk will vary between media formats (Adam, 1998: 167; Sjöberg and Wahlberg, 1997: 11). Certain media technologies will be better equipped to set the historical context of risk, others will choose to focus on visual dramatisation (Kitzinger and Reilly, 1997: 340). Differences in the structure, style and presentation of risk will deviate between media forms such as television, radio, newspapers and the internet (Mythen et al., 2000: 43). Television producers, for instance, tend to prefer risk stories that are visually stimulating and dramatic (Anderson, 1997: 121; Negrine, 1994: 121).[10] Whereas Beck virtually equates the media with broadcast journalism, we might reasonably expect non-news-based representations of risk to feature in day-to-day sense making around risk. As Ungar (1998: 42) posits, popular cultural representations of risk can serve as 'potent metaphors in public discourse'. Striking fictional representations of risk – for example, within films, novels or adverts – may be measured up against real-life incidents.[11] The impact of popular cultural artefacts on risk consciousness is an under-researched area and one which escapes the attentions of the risk society thesis. As figural regimes of signification become increasingly popular, the political dimensions of representation are being remodelled (see Clark, 1997; Thompson, 1997). Undoubtedly there is a growing preference among programmers for 'infotainment' and 'docusoap' programmes that use public testimony to present risk as spectacle and/or drama. Such forms of representation – which tend to trivialise and individualise risk in equal measure – only serve to cast further doubt on Beck's portrayal of the media as a generator of rational environmental concerns. After all, it is not just a question of *if* lay actors are able to speak, but also *how* they speak. As Cottle reasons:

> Ordinary voices are routinely accessed ... but rarely are they granted an opportunity to develop their arguments or points of view at length, much less directly confront and challenge political and expert authorities ... positioned by the news media to symbolise the 'human face'... these voices in fact rarely find an opportunity to advance rational claims – whether 'social' or 'scientific'. (Cottle, 2000: 29)

Of course, public expectations of risk coverage will differ according to media formats. Broadsheet journalists may be expected to provide historically accurate accounts of risk situations and depth and detail of reporting will be attributed greater value than sensationalistic 'scoops'. Thus, while reading a daily newspaper might well amount to 'an exercise in technology critique' (Beck, 1992: 116), meanings made will depend upon the politics of the paper one takes and its style of presentation. As will be illustrated in Chapter 5, public understandings of risk are influenced by access to and choice of media forms. Hence, although it is logical for research to focus upon dominant sources of public information, such as newspaper and television news coverage, it is illogical to address the media as a monolithic block, rather than a differentiated system (van Loon, 2000b: 234).

The media cannot realistically be perceived as acting in favour, or indeed against, the relations of definition in any uniform fashion. Unfortunately, Beck's tendency to 'lump the media together' leads to a one-dimensional understanding of media output (Anderson, 1997: 188). As Hansen's (1990; 1991, 2000) work indicates, we can expect to find cross-cultural variations in media output and audience preferences. Beck himself is neglectful of cultural diversity and falls short in appreciating that the selection and coverage of risk issues will reflect economic, geographical and cultural conditions (Linne and Hansen, 1990).[12] In opposition to the risk society approach, pressure to acquiesce in the values of power-holding institutions is ingrained in the media production process. News sourcing, news values and journalistic preferences will all shape the content and range of media output. These structural factors suggest that the reporting of risk does not take place on a level playing field. Rather than acting in direct opposition to the relations of definition, media professionals are likely to accept that the game has to be played on a slanted pitch.

READING THE MEDIA

In the preceding sections, it has been argued that the risk society thesis misses the political, economic and organisational processes which shape and influence media production. In the remainder of the chapter, I wish to press on to consider the issue of media consumption, through an analysis of the relationship between media representations and public understandings of risk. This inquiry will form the basis for a more general discussion of risk perception in Chapter 5.

In an evolving western techno-culture, public demand for risk information seems unrelenting. In view of the thin dividing line between alerting the public and creating undue panic, the mass media is saddled with a hefty weight of social responsibility. Despite the negative possibilities of generating moral panic, theorists such as Giddens (1999) argue that a degree of scaremongering about risk – be it intentional or otherwise – is justified in order to develop public awareness. Others, such as Anderson (1997: 167) and Reilly (1999:

131) are more reticent about the benefits of media amplification. Drawing upon ethnographic research, Reilly (1999) argues that routine exposure to mediated risks can encourage lay publics to become blasé about the possibility of personal danger. With reference to the first wave of media interest in the BSE crisis, a number of Reilly's respondents became so exasperated by the constant blizzard of information, they took to customarily switching channels to eschew further news. In the early stages of the crisis, respondents believed that the media coverage of the BSE crisis was sensationalist, considering that BSE was a subject that 'no-one, not even the experts really knew anything much about' (Reilly, 1999: 132). In Reilly's first study in 1992, almost half of the sample group actively rejected alternative information on BSE because of the possibility of media sensationalism (Reilly, 1999: 132). One respondent summed up the general mood by referring to the BSE crisis as, 'yet another media food scandal that we were all sick to the back teeth of' (Reilly, 1999: 131).[13] Right-wing commentators may take this desire to 'shut out' risk as an illustration of public naivety. We might, however, more usefully surmise that people are less likely to be concerned about risks when lived experience indicates that the probability of them being affected is remote. It is quite understandable that social actors will focus their energies on avoiding more prosaic risks, such as losing one's partner, one's job or one's house. Regardless of the level of media exposure, certain risks may be construed as beyond the range of influence of ordinary individuals.[14]

In contradiction to Beck's thesis, prolonged coverage of risk by the mass media does not ineluctably facilitate public reflexivity. Although certain studies have shown that the media has the capacity to act as a public watchdog in risk situations (Reilly, 1999), interpretations of media communications are vitally indexed to power and representation. For some, perceived powerlessness can lead to pragmatic and fatalistic interpretations and responses. For others, media amplification of risk may result in public scepticism about levels of harm (Anderson, 1997: 167). Rather than recognising that risk interpretations are manifested along a continuum, Beck's penchant for hyperbole periodically projects him into a position of media-centrism, in which cultural knowledge about risk is reduced down to representation: 'No mass media information, no consciousness of risk' (Beck, 1987: 155). Rallying against such indiscriminate claims, ethnographic studies indicate that local sources such as friends, family, work colleagues and health professionals are important wells of advice

(Caplan, 2000a; Reilly, 1999). The media acts as a stimulus within, rather than an end point in the risk communication process (Dunwoody and Peters, 1993). Even if we assume that the media *is* the most important single source of risk information, public consumption of media products remains diverse as opposed to uniform. Risk communications will produce variable affects, with individuals responding to risks in locally contingent ways (see Cottle, 1998: 21; Tulloch, 1999: 34; Tulloch and Lupton, 2001). As will be explained in Chapter 6, much will hang upon the environment in which media messages are encoded and the cultural context in which they are decoded. Material, cultural and geographical features are of vital significance in determining how media texts are read and reproduced (Hall, 1980; Lodziak, 1986; Morley, 1980). Resource-related factors such as educational access, scientific knowledge and technical familiarity influence the meanings made of risk information. Taking on board the complexities associated with media representations of risk, the most likely beneficiaries will be those who already possess background knowledge of the subject in question (Anderson, 1997: 200).

Painting over these important qualifications, Beck maintains that the media acts to challenge the dominant relations of definition, generating public pressure for stronger regulation of environmental risks. As lay actors gain expertise in risk issues, more intense forms of public consciousness develop. Yet the available empirical evidence negates the possibility of uniformly reflexive responses to risk information. Beck's portrayal of the media as a catalyst for public reflexivity does not venture far enough along the hermeneutic route and fails to explain how risk communications are translated into risk meanings. In short, the risk society thesis tramples over the active, communicative 'work' involved in the social construction of risk. The efficacy or otherwise of media messages will be influenced by embedded structural features and the situated nature of the audience. Beck's understanding of the public as an amorphous mass, waiting to be schooled in reflexivity is decidedly out of kilter with contemporary media and cultural theory. In the risk society, counter-experts, political dissenters and protesters are embraced by the public as the keepers of risk truths. Ironically, therefore, Beck assumes passivity on behalf of the same audience he wishes to credit with a critical and reflexive consciousness. It is almost as if the audience are attributed a critical attitude toward the relations of definition, but not to its opponents and detractors. These criticisms suggest a lack of attention to the way in which risks are interpreted by social actors in embedded cultural contexts. Often, Beck simply equates media *reporting* with media *effects*, demonstrating little regard for empirical studies of risk perception (Beck, 1987: 155). This omission leads him to undervalue the active role of the audience in decoding media representations (Lupton, 1999b: 7; Tulloch, 1999: 56). The moment of risk perception cannot possibly arrive prior to the interpretation of information and images. Whilst in the risk society thesis 'information equals reality', we would do well to heed Douglas' reminder that 'information does not even become information at all unless it is somehow coded by the perceiver' (1985: 27).

CONCLUSION

In this chapter we have confirmed that Beck's account of the role of the media in the social construction of risk fails to delve beneath the surface layer of representation. As a result, the fundamental issue of how individuals make sense of media products goes unplumbed. This oversight leaves the risk society thesis with an outmoded hypodermic model of media effects. In the end, Beck ends up in a cul-de-sac of media-centrism without having much of an idea about how he got there, or which is the best way out.

It is worth restating that the interpretative outcomes of risk communications are multifaceted. Thus, more sophisticated comparative research is required to establish concrete patterns within a presently scattered field (Sjöberg and Wahlberg, 1997). Further, it is evident that in situ ethnographic research provides the most effective means of gaining insight into the preferences and interpretations of media audiences (Sparks, 2003: 197). As far as the more expansive risk society argument is concerned, a horde of criticisms have come to light. First, Beck's theoretical impression of the process of risk communication lacks empirical support and is reliant on anecdotal evidence. Second, the risk society account of the relationship between the media and the public is contradictory and imprecise. Third, scant attention is paid to either the political economy of the mass media, or the routine features of the media production process. As has been demonstrated, the political, organisational and economic contexts in which media interests operate have vital implications for the reporting of risk. By divorcing his analysis from any particular cultural context, Beck is able to make the media mimic a Habermasian utopia; a platform for undistorted information and public debate. The present economic environment in which the media operates in the West throws such a viewpoint into sharp relief. It should be itemised that the continued expansion of the media is taking place in *private* hands, for *profit*, and *without* social regulation (Thompson, 1997). Inter alia, this questions the ability of media organisations to provide free and undistorted information about risk. Fourth, as we shall go on to explore in Chapter 5, Beck's work is characterised by a disregard for the cultural hermeneutic involved in structuring everyday under-standings of risk. This is particularly poignant with regards to the grounded habitus in which people interface with the media.

Stepping beyond the risk society debate, our discussion has raised several issues of wider consequence. What is at issue here is not only the way in which risks are routinely represented in the media, but also the degree and quality of information about risk available to lay actors. Public knowledge and subsequent behavioural intentions – be they preventative, combative or dismissive – depend upon a diverse and unrestricted stream of information. In order to make informed lifestyle choices, the public require accurate media information about risk. Of course, allowing for the volatile and unknown quality of manufactured risks, objectivity is an unattainable goal. Furthermore, the economic and organisational context out of which risk knowledge is disseminated questions the likelihood of due impartiality being realised in practice: 'Beck's voices of the "side effect" are all too often rendered socially silent, notwithstanding their statistical and symbolic news presence, and they remain the discursive prisoners of tightly controlled forms of news entry and representation' (Cottle, 2000: 43).

At present, the flow of information about risk is managed and controlled through implicit and explicit forms of censorship. The potential consequences of this for the formulation of public knowledge give rise to concern, particularly set against the prevailing power of global economic forces which continue to transform the media from a space of rational discourse to one of figural entertainment and spectacle (Boyne, 2003: 31; Lash, 1990: 174; Thompson, 1997: 35). This subterranean shift from discursive to figural regimes of signifi-cation has far-reaching implications for meaning making in the public sphere. Ultimately, it is only possible for audiences to be 'active' with the information they are able to access. Taken collectively, these criticisms fundamentally question whether the mass media has the potential to stimulate public reflexivity to the point of the effective oppositional action envisaged by Beck.

5
Perceiving Risk

Having reached the midway point of the book, it is worth taking an over-the-shoulder glance backwards at the themes discussed and a directive nod forwards toward the remaining issues to be dealt with. Hitherto, we have centred on the physical and social construction of risk; that is, the means by which risks are produced, defined and mediated in contemporary culture. Using the risk society thesis as a pivot, we have observed the various ways in which dominant institutions materially manufacture, technically assess and socially represent risk. In the second chapter, the environmental consequences of techno-scientific and industrial development were brought to the fore. In Chapter 3, we inspected the brief of law, science and government in overseeing and managing risk. We have also explored the distinctive institutional position of the mass media as a node of risk communication, explicating the organisational, economic and political factors which direct media output.

Henceforth, we will advance the debate by moving more decisively into the territory of risk consumption. This shift in gear necessitates a more focussed examination of the impacts of risk on the practices, perceptions and values of individuals within contemporary western cultures. The chapters left over will be driven by a desire to understand how risks are conceptualised, negotiated, countered and consumed within everyday life. In this chapter, we direct attention towards the cognitive aspects of the risk society thesis, applying Beck's metatheory to grounded empirical studies into risk perception. As a lead in to future discussions, we also scan the current relationship between experts and the public and comment on the import of cultural influences on lay understandings of risk. In Chapter 6, we rotate toward the substantive effects of risk and individualisation on social structures, chasing out the implications of changes in the family, workplace and personal relationships. In the penultimate chapter, we return to the cognitive thread, considering the extent to which the interface between risk perceptions, reflexivity and trust relations shapes interactions between experts and the public in contemporary society. Finally, in Chapter 8 we attend to the political movements and motions generated by the wider permeation of risk into the public sphere.

Following the pattern established in previous chapters, the risk society thesis will be set against the findings of empirical research. Through sustained analysis of existing studies, the lack of symmetry between theory and practice will be illuminated. We begin by recounting Beck's portable approach to risk perception, leaning upon the commentaries of Lupton (1999a; 1999b) and Culpitt (1999). Subsequently, we present an abridged summary of the findings of cognitive research into risk perception. As a corrective to the prevailing psychological wind, I offer up two outstanding cultural studies which in turn offer a more penetrative route into public understandings of risk. These contextual sections pave the way for a more informed comparison of the risk society thesis and empirical risk research. In evaluating the correspondence between the risk society perspective and empirical studies into risk perception, I address three core questions. First, what do we confidently know about public under-standings of risk? Second, can existing empirical research enable us to access the cultural dimensions of risk? Third, to what extent does the risk society thesis grasp the dynamics of the hermeneutic process?

THEORISING RISK PERCEPTION

Beck argues that, in contemporary society, public attitudes towards risk cannot be adequately conceptualised as individual fears about unavoidable external dangers. In the risk society, cultural understand-ings go beyond individualistic reasoning and simple attribution to fate. As recounted in Chapter 3, since the Enlightenment period, knowledge about risk has steadily developed within western cultures. Enhanced scientific and social knowledge has led to personal techniques of avoidance becoming customary within everyday life. Following this logic, Beck contends that contemporary protection strategies involve not only accounting for private dangers, but also thinking through the more public effects of risk. The encompassing span of global risks leads to widespread recognition of the unfavourable effects of modernisation on the lived environment and produces peculiar social and psychological effects. Socially, shared notions of safety and security become diluted (Beck, 2002). Psychologically, anxiety and insecurity become an integral part of the modern condition (Wilkinson, 2001: 4).

In order to get a handle on risk perception, Beck marks out two perspectives that have traditionally been used as framing mechanisms: 'natural objectivism' and 'cultural relativism' (1995a: 162). The natural objectivist approach, adhered to by the relations of definition, is based upon scientific knowledge and economic calculation. As discussed in Chapter 3, the natural objectivist model has dominated institutional risk-assessment practices, being widely employed within medicine, health, law, economics and engineering. Within these fields of inquiry, risks have been perceived as measurable phenomena to be identified, assessed and quantified. In treating risk as an objective and calculable entity, natural objectivism is broadly compatible with the realist position (Beck, 1999; Lupton, 1999a: 33).

In stark contrast to natural objectivism, cultural relativism suggests that the meaning of risk cannot be objectively determined. Rather, risk is deemed to be a social reality constructed via the reproduction of shared ideas and values. For relativists, perceptions of risk are culturally formed as a result of the interplay between institutional discourses and individual subjectivities. Thus, relativists posit that risks are inseparable from cultural belief systems and cannot be meaningfully objectivised (Dean, 1999). As a result, cultural relativism has been closely aligned with social constructionism within the social sciences (Lupton, 1999a: 60).[1] In *Ecological Politics in an Age of Risk* (1995a) Beck contends that objectivism and relativism each have particular merits and shortcomings as methods of perceiving risk. Natural objectivism has been assisted by the application of technology and monitoring procedures that have advanced the quantification of risks. As Beck (1999: 23) points out, the objectivist method has enabled scientists to identify and quantify environmental dangers, such as the hole in the ozone layer and the appearance of acid rain. However, on the minus side, the objectivist approach assumes that risks are extraneous entities and does not entertain either the social production, or the cultural cognition of risk. Beck (1995a: 90) tells us that technical experts employing the paradigm have tended to see it as a panacea and have failed to acknowledge that objectivism is ultimately a value position, not an immutable truth.

Acting as an opposite, relativism avoids the detached approach of objectivism by taking account of the culturally situated character of risk cognisance. However, Beck believes that stiff cultural relativists fall short in distinguishing between degrees of impact and block out the distinction between natural and anthropogenic risks (Beck, 1999: 23). Hence, the relativist approach to risk tends to lose sight of 'objective' degrees of danger and the 'special features of large scale technological hazards' (Beck, 1995a: 162). Of course, strict application of cultural relativism would lead to the collapse of Beck's historical scheme, blurring epochal boundaries and threatening the differentiation between natural hazards and manufactured risks. Although the possibilities of natural objectivism and cultural relativism are frequently explored in the risk society narrative (1995a; 1996a; 1999), Beck does not religiously adhere to either an objectivist or a relativist position (van Loon, 2000a: 176). As Lupton notes:

> He maintains a 'natural-scientific objectivist' approach by subscribing to the idea that 'real' risks exist, but brings in 'cultural relativism' by arguing that the nature and causes of risks are conceptualized and dealt with differently in contemporary western societies compared with previous eras. (Lupton, 1999a: 61)

In theorising risk perception, Beck advocates a 'sociological perspective' which draws upon the finer points of each approach (Beck, 1995a: 76). In theory at least, by marrying realism to constructionism the sociological perspective is free to explore both the concrete and the abstract dimensions, imbuing risks with an objective reality whilst also differentiating between cognitive effects:

> I consider realism and constructionism to be neither an either-or option, nor a mere matter of belief. We should not have to swear allegiance to any particular view or theoretical perspective. The decision whether to take a realist or a constructionist approach is for me a rather *pragmatic* one, a matter of choosing the appropriate means for a desired goal. (Beck, 2000d: 211)

This elastic approach is stretched over the risk society thesis and manipulated to suit the argument. As we saw in Chapter 2, Beck's understanding of environmental risks follows the tradition of natural objectivism, with scientific evidence of ecological demise being unreservedly accepted (Goldblatt, 1995: 174). By contrast, a relativist stance underpins the evaluation of the dynamics of contemporary interpersonal relationships (Beck and Beck-Gernsheim, 1995: 52). In relativistic mode, Beck is receptive to heterogeneous responses to the 'normal chaos' of loving partnerships.

Despite affording theoretical manoeuvrability, Beck's untethered approach also generates contradictions. In shifting between realism and relativism, the angles of the risk society thesis become somewhat abstruse (Alexander and Smith, 1996: 251). All in all, Beck is unclear about whether manufactured risks are extant or imagined: 'risks are a kind of virtual, yet real, reality' (Beck, 1998b: 11). As described in Chapter 4, on occasion, Beck is adamant that manufactured risks are imperceptible to human faculties and depend upon scientific verification: 'our senses have become useless' (Beck, 1987: 155).[2] In this groove, contemporary risks evade both identification and perception: 'In matters of risk we have been disenfranchised ... we the citizens have lost sovereignty over our senses, and thus the residual sovereignty over our judgement' (Beck, 1987: 156).

In such instances, Beck follows a 'top-down' model of risk perception, inclining toward natural objectivism. However, at other moments, Beck treads much closer to the relativist position, stressing the culturally situated nature of risk perceptions (Beck, 1996a: 3; 1999: 22). In this mode, Beck is appreciative of the socially constructed and diverse nature of public understandings of risk: 'the same dangers appear to one person as dragons, and to another as earthworms' (Beck, 1999: 22). Following relativism, the suggested model of risk perception here is multilinear, structured by social context rather than expert identification. Even though the risk society thesis borrows from both realism and relativism – as with Beck's positioning of the media – measurably different weights are placed on each approach. Taken in its totality, the risk society perspective is predominantly informed by a realist rather than a relativist position (see Alexander and Smith, 1996; Lash, 2000: 51). At best, Beck's work draws upon a 'weak' form of social constructionism, compared with the stronger versions present in anthropological and governmentality approaches (Lupton, 1999a: 29). As will be detailed in Chapter 8, followers of Foucault contend that risks are constituted by discursive practices and are unrecognisable outside constructed belief systems. Residing some distance from this position, Beck (1999: 23) avers that risks are objective, hazardous and deleterious, regardless of cultural beliefs and values. Such an objectivist approach to risk has prompted several critics to accuse Beck of artificially separating out public and scientific knowledge about risk (Dickens, 1996; Wynne, 1996). For Beck, there are two sides to understanding hazards: 'the risk itself and public perception of it' (Beck, 1992: 55). Hence, clear water is placed between existing 'objective' facts and 'subjective' values. As Dickens (1996: 40) notes, this heavy-handed separation pushes Beck's construal of public understandings of risk perilously close to the expert bodies he chooses to criticise.[3] Thus, a curious paradox can be identified at the hub of the risk society thesis. Having vigorously criticised expert bodies for their use of the objectivist paradigm, Beck proceeds to adhere to a universal model, insisting on the objective existence of risks. In order to retain generality, Beck is obliged to assume that manufactured risks are objectively 'out there' and potentially life-threatening. Furthermore, the universal thrust of the argument decrees that risk perceptions are regimentally ordered: 'every (risk) event arouses memories of all the other ones, not only in Germany, but all over the world' (Beck, 1996a: 114). As we shall see, a strong gravitation towards objectivism rather colours Beck's understanding of public perceptions of risk.

RESEARCHING RISK PERCEPTION

Having sketched out Beck's transportable approach to risk perception, we are now in a position to relate the risk society thesis to existing empirical work. However, prior to establishing the fit between Beck's theoretical treatise and grounded studies, it is first necessary to provide a nuts-and-bolts account of the overall findings of research into public perceptions of risk.[4] In exploring the intersection between qualitative inquiries into risk perception and Beck's risk society thesis I also intend to flag several methodological concerns which arise out of the dominant tradition of empirical research.

So, what kinds of research have been undertaken in the field of risk perception? What have been the most prominent findings of research into public understandings of risk? Historically, the bulk of empirically based risk research has been conducted in the United States (Douglas, 1985: 8; Krimsky and Golding, 1992). The majority of American research studies have adhered to the cognitive-scientific perspective, comparing individual perceptions of risk with statistical probabilities of harm. Empirical research has also sought to probe behavioural intentions in hypothetical situations and assessed the psychology of the decision making process (see Slovic, 2000). The prevailing research methodology utilised in such studies has been psychometric testing, through which researchers have attempted to identify various cognitive strategies (Flynn et al., 1994; Slovic 1987; 1992). Through cognitive-scientific studies of risk perception it has been established that individuals use certain heuristics and biases in order to construct understandings of risk incidents (see Joffe, 1999:

56). For example, most people possess an 'optimistic bias' which leads them to underestimate the probability of being adversely affected by risk (Weinstein, 1987). Psychometric studies into risk perception have indicated that individuals feel an unjustified sense of immunity with regards to risks that arise from familiar activities (Lee, 1981; Slovic et al., 1981, Slovic, 1987). In what is considered to be a seminal study, Slovic et al. (1981) report that individuals generally overestimate the risk presented by rare but memorable events, whilst underestimating the threat posed by more mundane risks. Slovic et al. (1981) also detail that risk events which cluster together are perceived to be more serious than one-off events. In addition, disasters that erupt immediately are more likely to provoke anxiety than those which are temporally staggered (Slovic, 1992).

The formation of risk consciousness must also be understood in relation to the operation of group dynamics (Joffe, 1999: 31). Empirical studies have consistently found that groups will tend to make riskier decisions than individuals (see Douglas, 1985: 58). By resorting to 'group-think' the responsibility for risk is shared by the collective and the individual burden is lessened (Dion et al., 1971). At a broader level, cultural groups are quite adept at attributing responsibility away from themselves and towards others. Sooner than facing risk in a socially responsible fashion, both lay and expert actors may succumb to the temptation of designating blame. As Joffe (1999: 34) points out, at a psychological level, the 'not me – other' approach is a handy way of absolving personal culpability and despatching blame toward targeted groups.[5]

A frequently reported finding within cognitive risk research is that the degree of individual choice involved will affect public attitudes to risk exposure. Risks that allow a high degree of agency have been found to be less objectionable than risks which are perceived to be visited upon individuals without due consent. It would seem that 'taking' a risk is of a completely different cognitive order than being 'subjected' to one, even if the two poles are rarely clear-cut. What is significant is not whether the risk is voluntary or imposed, but whether it is *perceived* to be so. Cognitive effects are not necessarily integral to the risk itself, rather they are the products of imagined outcomes (Bennett, 1998: 6). Psychometric studies have shown that harmful types of risk that have the capacity to produce serious injury or death – for example, nuclear radiation, murder and asbestos poisoning – are likely to evoke 'dread' (see Slovic, 2000).

In the British context, Claire Marris and Ian Langford (1996) have extended the findings of American psychometric research, exploring public attitudes to particular risk incidents. Using a sample of 210 respondents, Marris and Langford (1996: 36) sought to test the hypothesis that global dangers are universally feared, whilst local risks are more easily tolerated. In order to investigate this proposition, a mix of 13 risks – including terrorism, sunbathing, genetic engineering and alcohol consumption – were considered, with respondents being invited to rate the seriousness of each.[6] The particular selection of incidents was deliberate, with the researchers claiming that the risks were classifiable by two variables. The first variable was the extent to which the harmful effects of the risk might be delayed and catastrophic. The second, the extent to which the risk was imposed or voluntary (Marris and Langford, 1996: 36). Concurring with Slovic (1992), Marris and Langford confirm that familiar voluntary hazards, such as microwave-oven usage and alcohol consumption are perceived to be low-risk activities. Meanwhile, global dangers such as genetic engineering, ozone depletion and nuclear power were rated as highly risky and approached with a general sense of 'dread' (Marris and Langford, 1996: 36).

Marris and Langford (1996) were also interested in the ways in which personal 'worldviews' tailor individual perceptions of risk. Exploring the personality types proffered by Mary Douglas and Aaron Wildavsky (1982), Marris and Langford identified 'a remarkable consistency of ideas expressed by people of the same cultural disposition' (Marris and Langford, 1996: 37). This indicates that understandings of risk will be canalised according to political outlook and personally held values. Indeed, the heterogeneous nature of public perceptions of risk is axiomatic within risk research, being reinforced by a plethora of empirical inquiries (see Finucane et al., 2000; Flynn et al., 1994; Slovic 1993).

Various research studies have demonstrated that cultural factors such as class, gender, age, and ethnicity will shape understandings of risk. Evidence that perceptions of risk are influenced by social class is provided by Graham and Clemente (1996), who report that men with higher educational qualifications and higher incomes are less risk-averse than other groups. The apparent class factor raises tricky ethical issues and problematises Beck's depiction of the uniformly risk-averse individual. As Douglas (1985: 21) notes, a blue-collar worker whose plant job is at stake might be in favour of nuclear power, whereas middle-class elites concerned with preserving their mountain holidays may be firmly against the production of nuclear energy. Flanking class, a number of research studies have indicated that gender plays an important role in structuring perceptions of risk. Ceteris paribus, not only will males and females be troubled by different risks, they will also perceive the same risks in a dissimilar manner (Gustafson, 1998). Using psychometric testing across a range of risks, Flynn et al. (1994) contend that men are generally less anxious about risk than women. From a sociological angle, others have argued that women's comparatively developed sense of risk awareness may arise out of differential patterns of socialisation (Rose, 1993: 143). In western cultures, women have traditionally been socialised into risk awareness in the areas of personal safety, hygiene and sexual health (Weaver et al., 2000: 172). By contrast, males have historically been encouraged to be more fearless and to actively engage in risk-taking behaviour (Douglas, 1985: 70; Walklate, 1997). Situated in this context, it is unsurprising that research shows that women tend to use a greater range of safety strategies and precautions than do men (see Gardner, 1995; Stanko, 1996).[7]

Age is also thought to have an important bearing on attitudes towards risk (Field and Schreer, 2000; Hinchcliffe, 2000: 127). Empirical studies consistently demonstrate that elderly people tend to overestimate the possibilities of danger and are more likely to feel threatened than younger people (Balkin, 1979; Mooney et al., 2000). Naturally, children's perceptions of risk will vary substantially from those constructed by their parents (Furedi, 1997: 117; Jackson and Scott, 1999; Scott et al., 1998). Although the structuring influence of ethnicity on risk interpretation appears to be an under researched area, studies have noted that different ethnic groups will fashion particular attitudes towards risk (Caplan, 2000a; Mackey, 1999). Having studying a variety of ethnic cultures in America, Finucane et al. (2000) found that white people were generally less anxious about a range of risks to health than were people of colour. Hence, both the modelling of risk incidents and interpretations of risk communications are likely to be affected by ethnic identity (see Burger et al., 1999; Flynn et al., 1994).

CHASING OUT THE METHODOLOGICAL LIMITATIONS

Before pitching the main findings of empirical research against the risk society perspective, there is merit in documenting the methodological problems that arise out of the framework employed in cognitive-scientific studies. In particular, the actual design of psychometric research reproduces a realist understanding of risk and embodies a rationalist interpretation of human motivations. The American tradition of psychometrics has tended to view participants, (mis)perceptions through the eyes of the informed and knowledge-able researcher. However, a fixed idea of the individual as an instrumental and logical decision maker is based on the prototype of the rational investigator: 'both are driven to seek order in the world; both recognise inconsistency; both assess probability' (Douglas, 1985: 28). Thus, the 'mirror-imaging' of lay participants with expert researchers may have had the effect of routing American research along the path of a one-dimensional rational-choice model of risk perception. This unwavering focus upon the subject as an unemotional rational actor silences the embedded cultural factors which shape understandings of risk.

A related criticism is associated with the predilection for individ-ualising and privatising public perceptions and motivations. Cognitive-scientific studies have tended to conceive of risks as potential threats to the isolated individual, as opposed to collective dangers faced by society. This individualistic bent has been exacerbated by a narrow research design and a decontextualised methodology. The use of psychometric testing under laboratory conditions and/or one-on-one interviews in neutral settings encourages cognitive-scientific studies to treat participants as disconnected individuals rather than culturally joined-up actors. As such, cognitive studies may actually be designed in a way which assumes that individuals rather than populations carry social risks. Running against such pre-conceptions, the general threat presented by many risks to public health means that hazards can no longer be adequately conceptu-alised in private terms (Prior et al., 2000: 106). Perceptions of risk are constructed through communicative exchanges with significant others, associates and expert institutions (Caplan, 2000a: 23). Therefore, 'public' understandings of risk cannot be adequately realised by measuring individual responses to particular hazards. As we have seen, attitudes towards risk are socially constructed and housed within collective cultural networks (Lupton, 1999b: 15). Unfortunately, the lion's share of psychometric studies have remained constrained within the neat boundaries of the cognitive-scientific model and have followed the orderly theory of rational choice. As a result, an overly rigid, value-rational understanding of human behaviour has been reproduced. Sadly, the paradigm that dominated risk research in the twentieth century may not be especially helpful in enabling us to understand public responses to environmental risks in the twenty-first century.

It is important to acknowledge that risks – and, by dint of this, responses to risk – are protean. Despite retaining usefulness as pointers toward the average, psychometric studies into public perceptions of risk only work in broad descriptive categories. Furthermore, too much has been made of 'errors of judgement' and 'heuristics and biases' which influence public perceptions and not enough of the social and cultural factors which shape interpretations of risk. We will return to this aspect in Chapter 7, demonstrating that differences between 'objective' experts and 'subjective' lay actors are not indicative of the latter's inability to assess the probabilities of danger. Remaining sensitive to the crudity of binary distinctions, it is expectable that a divergent logic may emerge between 'expert' and 'lay' approxima-tions to risk (Bradbury, 1989; Taig, 1999). Needless to say, these styles of sense making are liable to conflict, but this does not mean that one logic should be seen as superior to another.

In reproducing the expert/objective and lay/subjective couplets, cognitive-scientific studies of risk perception have trampled over the intricate dynamics involved in the formulation of public understand-ings of risk. The source of this myopia stems from an over-eager endorsement of the objectivist perspective. For this reason, several empirical studies have simply failed to account for the role of collective networks and symbolic factors in the formulation of risk perceptions (Lash, 1993; Lupton, 1999a: 30). Cognitively based studies may dif-ferentiate between risks, but they 'fail to incorporate adequately wider social and political contexts in which risks and benefits come to be evaluated by individuals' (CSEC Report, 1997: 4). Under controlled conditions and outside of an everyday social context, respondents may feel compelled to make sense of risk using analytical techniques as opposed to habitual anchors. This suggests that the research framework employed in cognitive-scientific studies limits the horizons of interpretation, even before lay voices are permitted to speak. It must be recognised that individuals tend to encounter everyday risks with a pre-existent package of beliefs and assumptions (Douglas, 1992: 58). As we shall see, attitudes to risk are indelibly cultured and will be formulated by emotional as well as rational referents (Lupton, 1999a: 30).

DEVELOPING A CULTURED APPROACH

In the remaining chapters of the book, we will continue to chew over the salience of culture in nourishing public understandings of risk. In order to provide a makeshift compass, in this section we spotlight two noteworthy cultural studies which serve as useful correctives to cognitive-scientific approaches to risk perception. Rather than extracting participants from their everyday context, the socio-culturally informed inquiries undertaken by Macgill (1989) and Reilly (1999) utilised valuable ethnographic methods to draw out the myriad of meanings attached to risk in everyday life. Both studies sought to access the situated context of sense making, concentrating on the way in which actors interpret risk within lived cultural environments.

The purpose of Macgill's (1989) inquiry was to gauge lay responses to the risk of radioactive discharge from the Sellafield nuclear plant in Britain. In the light of abnormally high rates of childhood leukaemia in the area, Macgill (1989) chose to interview a cross-section of residents living within close proximity of the nuclear plant. Using ethnographic research, a wide range of opinions and standpoints on the risk of radioactive contamination were assembled. The diversity of risk perspectives were contrary to both conventional 'expert opinion' and national media coverage, both of which gravitated towards absolute positions on risk. At the time, dominant media representations were suggestive of a united culture of local opposition to the plant. In actuality, many interviewees flatly denied any possible link between health risks and the nuclear power plant (Macgill, 1989: 58). Through Macgill's fieldwork the inherent complexities of lay responses are brought to the surface, emphasising that risk perceptions are socially variable and culturally situated.[8] Public understandings of risk are not simply conditioned responses to knowledge from above. Lay actors accumulate, assess and disseminate risk information over time, meaning that perceptions of risk will always be culturally contingent. As the Sellafield study demonstrates, cultural readings of risk are moulded by a plethora of factors, such as social status, economic factors, collective networks and mass media representation (Macgill, 1989: 55).

A decade after Macgill's study, Jacquie Reilly (1999) published her longitudinal research into public understandings of the BSE crisis in Scotland. Reilly's findings were drawn from two research projects seeking to investigate the production and reception of media representations of risk. As a valuable corrective to psychometric research,

Reilly attempted to access the collective dynamic of interpretation by selecting interviewees who were socially affiliated. The situated character of the research was enhanced by utilising group discussion as opposed to isolated interviews. By staggering the two projects – the first study took place in 1992, the second in 1996 – Reilly was able to tap into the dynamism of public perceptions of risk and to grasp the ways in which lay understandings evolved over time (Reilly, 1999: 129). In contradistinction to the cognitive-scientific studies, Reilly's research recognises the active role of individuals. Far from passively waiting for information about vCJD to be filtered down from experts to the public, many interviewees had sought to proactively gather information and to decide for themselves:

> Respondents used media coverage as a starting point for information which led to them phoning doctors and health authorities for information, picking up leaflets in supermarkets and butchers, asking shopkeepers where their meat came from, avoiding establishments such as cafés and restaurants which did not have clear signs/information about BSE and, in some cases, phoning restaurants to find out the source of their meat before booking. (Reilly, 1999: 134)

In sharp contrast to the cognitive-scientific approach, Reilly's work aptly demonstrates that the relationship between lay publics and experts is symbiotic rather than unilinear. As discussed in Chapter 3, with the benefit of hindsight, public alarm about the spread of BSE through the food chain proved to be justified. Presented as a couplet, Reilly and Macgill's studies afford much needed access to the cultural dimensions of risk. Deviating from the cognitive-scientific norm, both researchers develop situated approaches which are attuned to the culturally embedded nature of perceptions of risk.

THE RISK SOCIETY THESIS AND EMPIRICAL RESEARCH: AN UNCOMFORTABLE MARRIAGE?

So far in the chapter, we have outlined Beck's position on public perceptions of risk and presented the principal findings of empirical studies. We have also touched upon the methodological shortcomings common to cognitive-scientific approaches and hinted at the value of cultural studies in enabling us to understand public attitudes towards risk. We will return to underscore the value of culturally sensitive methods of risk research in Chapter 7. Having built up a clearer picture of the dynamics of risk perception, we are now well armed to relate the findings of empirical research to the broader curve of the risk society thesis.

Unquestionably, both cognitive-scientific and cultural studies provide some support for the risk society perspective. As far as cultural studies are concerned, Reilly's research suggests that lay publics are becoming less reliant on expert systems and are increasingly reflexive in their monitoring of social risks. Having recognised the contingency of 'official advice', participants in Reilly's study became actively involved in research and debate about vCJD. In line with Beck's suggestion of ailing trust in expertise, the government U-turn on BSE encouraged lay actors to re-evaluate their opinions and to entertain anti-establishment ideas (Reilly, 1999: 134). At a first glance, both Reilly's and Macgill's research lends weight to the risk society evocation of a developing culture of public reflexivity. Although we will mull over the question of reflexivity in greater depth in Chapter 7, it seems reasonable to assert that individuals in the West have acquired the habit of horizon scanning for risk and that nascent forms of reflexivity may be moulding around issues of risk.

In spite of being blighted by methodological deficiencies, several findings from cognitive-scientific research dovetail with the risk society thesis. Vindicating Beck's argument, cognitive studies have picked up on heightened public awareness of risk. More particularly, the 'icons of destruction' of the risk society – environmental pollution, nuclear technology and genetic engineering – have been frequently cited as sources of public concern (see Slovic, 1992; 2000). Empirical studies have also established that the shared characteristics of manufactured risks – dread, scientific uncertainty, unfamiliarity, voluntariness and irreversibility – will affect the perceived riskiness of situations and incidents (Bennett, 1998; Slovic, 1987: 280). At a methodological level, this moves us further along in understanding why the types of risks alluded to in the risk society thesis generate public anxiety. Nevertheless, it would appear that psychometric research has erroneously equated risk severity with risk anxiety. Genetic, environmental and nuclear catastrophes might well be the most feared risks – in psychometric parlance, the risks we 'dread'. Nonetheless, it does not follow that these global threats are the most cognitively consuming risks. Given their catastrophic potential, it is unsurprising that nuclear and genetic technologies are rated amongst the most dangerous risks by the public. However, the most *feared* risks are not necessarily the most *focussed upon* risks. If a person is asked whether the threat of nuclear catastrophe is 'more serious' than that of losing their job, they might reasonably answer in the affirmative. However, this belies the fact that one is likely to dedicate more time and effort to considering the detrimental effects of the latter than the former. Catastrophic risks have commonly been classified as the most anxiety-provoking risks, but they are not necessarily those which are most frequently ruminated on amidst the undulations of everyday life. It is probable that lay actors will actively distinguish between profane and catastrophic risks, responding to each in a qualitatively different fashion. As Anthony Giddens points out:

> Global risks have become such an acknowledged aspect of modern institutions that on the level of day-to-day behaviour, no one gives much thought to how potential global disasters can be avoided. Most people shut them out of their lives and concentrate their activities on privatised 'survival strategies'. (Giddens, 1991: 171)

In a similar vein, Culpitt (1999: 136) points toward 'two transparent and interlocking palimpsests of risk ... which are flung over each other'. The first is oriented to the personal and determined by private assessments of possible risks, the second is public-focussed and reflects concerns about unmanageable social risks. Although the seriousness of public risks may be widely acknowledged by lay actors, in everyday practice, people will be inclined to direct energies toward cognitively graspable and materially malleable issues:

> The scale of the risks that have burst the 'lifeworld' are so apocalyptic they can only be defended against – not easily resolved. The result of this is to alter the politics of responsibility so that, increasingly, individuals cannot be held responsible for the moral management of risks outside the area of the personal palimpsest ... faced with the massive encroachment of global risk we are forced to direct our resistance to 'abstractions' of the lifeworld ... the public palimpsest is so overwhelming that we are forced to return, almost atavistically, to inscribing the personal. (Culpitt, 1999: 137)

Following Culpitt, it seems reasonable to argue that, for the majority, the immediacy of risks within the everyday lifeworld will take cognitive precedence over potentially catastrophic but distant risks.

To cede this is not to argue against a wider trend of heightened public awareness of environmental risks. On the contrary, it is probable that individuals in western cultures are more informed about risks and hazards than at any other point in human history (Macionis and Plummer, 1998: 646). As the connections between the global and the local become more transparent, we should expect public understandings of risk to become even more sophisticated in the future (Boden, 2000: 193). Echoing the findings of Chapter 2, global environmental problems may be perceived as intangible items, beyond the sphere of personal influence. Against the risk society thesis, it is probable that individuals in western cultures will be motivated to act upon close-up personal risks rather than catastrophic global dangers. If, as Giddens (1991: 183) puts it, 'apocalypse has become banal', high-consequence risks may not permanently settle at the forefront of consciousness. Of course, we must resist artificially separating out 'local' actions from 'global' effects. Yet it remains the case that local and personal issues will tend to take precedence over abstract consideration of global affairs (Bennett, 1998: 14). This is perfectly understandable, particularly if previous actions have resulted in tangible securities, such as a stable relationship or greater job stability. The seemingly ubiquitous nature of uncertainty in contemporary culture means that people are obliged to divert their attentions towards selected environmental risks at particular times (Ravetz et al., 1989: 135). In the maelstrom of modernity, risks will invariably be dealt with on a 'first come, first served' basis. Thus, both risk perceptions and cognitive preoccupations will be changeable and scattered. As Beck avers, the dissemination of risk information by the mass media has unquestionably renovated public risk consciousness over the last half a century. Nevertheless, as was exposed in Chapter 4, the relationship between media representation and public consciousness is far from linear. Whilst news reports may draw upon expert knowledge to construct advice on strategies of risk avoidance, media products are themselves interpreted within culturally specific surroundings: the home, the pub, the workplace. Therefore, the sense which is made from risk communications will always be informed by proximate exchanges, as well as pre-existing values and experiences. This again reminds us that risk perceptions are socially and culturally constructed entities that cannot be properly be interpreted outside of frameworks of everyday lived experience (Dean, 1999; Douglas, 1985; 1992; Hinchcliffe, 2000). As will be revealed in Chapter 6, the development of public risk consciousness is intimately connected to a parallel trend of lifestyle choice and self-monitoring.

In spite of indicating an overall rise in risk consciousness, the empirical evidence is difficult to codify and does not neatly tessellate with the risk society thesis. Some of the research findings validate Beck's understanding of risk perception in contemporary society, others contradict it. The optimistic bias identified through empirical work illustrates that individuals are prone to underestimating the threat of proximate risks and exaggerating the scope of personal control (see Bennett, 1998: 9; Taylor and Brown, 1988). In many respects, neglecting familiar hazards – driving, DIY, smoking – is an eminently reasonable strategy. As Douglas (1985: 30) reminds us, attending to all the probabilities of disaster would lead to inertia. Lay actors will tend to prioritise their cognitive engagements with risk as a way of fending off feelings of engulfment. This rather tempers Beck's assumption that risk is increasingly eating its way into cultural experiences and colonising our cognitive maps.

In addition, studies into the heuristics and biases that affect personal risk judgements do not resemble the rational and reflexive subject recounted in the risk society thesis. As far as Beck depicts a reflexive and risk-observant individual, studies have consistently shown that perceptions of risk are commonly open to distortion and error (see Taylor and Brown, 1998: Weinstein, 1987). Public underestimation of proximate risks has led some theorists to argue that the 'real' risks are not the ones we fear the most (Furedi, 1997: 6; Marris and Langford, 1996). However, on reflection, such a proposition is rather difficult to substantiate. The fact that a particular risk may not directly impact on an individual who fears it does not invalidate its status as a risk. This is particularly pertinent in the case of 'unknown' public dangers, such as Aids, global warming or biochemical warfare. Since the deleterious effects of many environmental risks are rarely manifested instantaneously, the usual rules of quantification have to be suspended. As far as global threats are concerned, it would seem fairly fatuous to argue that an individual has 'more' – or indeed 'less' – chance of being affected by one risk than another. Unfortunately, several theorists critical of Beck's approach have failed to recognise significant qualitative variations in the composition of risks. For example, Furedi (1997: 23) insensately maintains that the threat to public health posed by global risks such as Aids has been vastly exaggerated. Drawing upon statistical evidence, Furedi contends that less remarkable accidents such as car crashes present a greater risk to public health than Aids. Simply flashing an eye over the global situation highlights the insouciance of such sentiments. Over 35 million people are currently infected with the human immuno-deficiency virus, with 15 million people already deceased through Aids.[9] The number of people dying of Aids per annum is rising exponentially, currently touching 2.6 million. In Britain, a recent report on Aids suggests that there are currently 10,000 people who are unaware that they are HIV positive, with 30,000 infected overall. Provided these statistics are accurate, roughly 1 in 2,000 people in Britain are already HIV positive.[10] Since manufactured risks such as Aids are highly latent *and* extremely mobile, the actual risk to the individual/society is impossible to predict. The greater the number of people that contract Aids over time, the greater the percentage chance of becoming a victim in the future. This recursive process is set to continue ad infinitum, unless a suitable cure for the disease can be developed and disseminated en masse.

Research into the effects of collective identities and value formation on mental models of danger also presents sticky issues for the risk society thesis, particularly at the level of personal responsibility. If, as the research suggests, there is a general tendency to attribute blame for risk to certain groups – Joffe's (1999) 'not me – other' phenomenon – then it would only be sage to question the depth of public reflexivity engendered by risk. If responsibility for collectively produced risk can be conveniently projected onto identifiable groups – the poor, homosexuals, the homeless, people of colour – its capacity to act as a progressive political lever is significantly lessened. In the case of Aids, media vilification of homosexuals for producing a 'gay disease' certainly helped foster complacency amongst heterosexual groups about the possibility of contracting the virus. Parenthetically, half of the 3,400 people diagnosed with HIV in Britain will have contracted the virus in 2001 through heterosexual sex as compared with just one fifth in 1991.[11]

Similarly, widespread recognition of the collective damage inflicted on the environment has not significantly altered patterns of production and consumption. Certain modifications have been made to consumer behaviour in the West, such as recycling and purchasing environmentally friendly products. However, the dominant system of capitalist production and consumption continues to expand in geography and magnitude. The feeling that we are all to some degree guilty of environmental pollution may serve to psychologically alleviate individual culpability for the manufacture of toxins and pollutants. As detected in Chapter 2, responsibility for environmental risk may be leaked through cognitive escape routes, causing a haemorrhage between risk recognition and everyday practices of precaution and regulation.

Thus, one of the potential pitfalls of polemical theory building is night blindness to the diversity and richness of lived experience. Unfortunately, the universal portrayal of risk perception in the risk society argument is insensitive to social and geographical variations (Elliott, 2002: 300). As will be demonstrated, Beck is guilty of exaggerating the uniformity of public perceptions and failing to account for the cultural dimensions of risk cognisance: 'complex issues of interpretation and meaning are swept under the carpet by the objectivist fallacy' (Alexander and Smith, 1996: 253). The inherent weaknesses within Beck's theory of risk perception can be attributed to a staunch refusal to engage with empirical evidence (Hajer and Kesselring, 1999; Smith et al., 1997). Beck resides exclusively in the theoretical ether and makes no attempt to visit existing research into risk perception. However, bold claims about public understandings of risk need to make researchable sense. Ruminating on the risk society thesis, there is more than a suspicion that Beck's own reflexive and risk-aware consciousness is being projected onto the population at large. Unfortunately, the empirical dimensions of risk perception are circumvented, threatening the wider credibility of the argument:

> It is simply not acceptable to assume that the empirical case has been made for the widespread existence of the increasing threat of risks and increasing risk perception, or that their combined impact on social behaviour and beliefs is so conclusive that we can properly herald the emergence of a new type of society. (Goldblatt, 1995: 174)

On the evidence gleaned from empirical investigation, perceptions of risk are less legible than Beck imagines. Contrary to the uniformity of public understandings alluded to in the risk society thesis, empirical studies illustrate that perceptions of risk are culturally canalised (Caplan, 2000a; Flynn et al., 1994; Graham and Clemente, 1996; Maguire, 1997). A variety of factors, such as class, ethnicity, gender and age have been found to affect public understandings of risk (Hinchcliffe, 2000: 127; Jackson and Scott, 1999: 102; Lupton and Tulloch, 2002b: 332). Which incidents and practices will be considered 'risky' differs according to cultural grouping and social affiliation

(Crook, 1999: 174; Lupton, 1999a: 15; Wilkinson, 2001: 17). For his part, Beck describes an even pattern of risk cognition, which glosses over the differentiated nature of public perceptions (Alexander, 1996; Skinner, 2000: 161). As Scott notes: 'The universalising language used by Beck is not sufficiently context-sensitive ... and the "we are all in the same boat" rhetoric distracts attention from differences both in exposure to and perception of risk' (Scott, 2000: 42).[12]

By relying upon an objective model of risk cognition, Beck neglects the structuring force of power relations in the formulation of risk knowledge. As will be discussed in Chapter 6, forms of cultural strat-ification are allied to the differential distribution of resources. Those with the least resources/power in society also tend to be the most heavily encumbered by the burden of risk (Day, 2000: 51; Hinchcliffe, 2000: 127). As Douglas (1985: 6) notes, society regularly 'exposes a large percentage of its population to much higher risks than the fortunate ten percent'. Hence, stratified risk experiences are cemented to the unequal distribution of resources in society. By consequence, the degree of exposure to risk and access to tools for reflection are key factors in the development of 'subjective' perceptions of risk. These conditions suggest that Beck's marriage between risk exposure and personal reflexivity cannot be taken for granted: individual subjectivity and self-reflexivity are resource-related entities:

> The self-reflexive individual as presented by Beck and Giddens, is a socially and economically privileged person who has the cultural and material resources to engage in self-inspection. Many people, however, simply lack the resources and techniques with which to engage in the project of self-reflexivity. (Lupton, 1999a: 114)

From this we can infer that education, social status and access to material resources will influence personal interpretations of risk. Adequate appreciation of the cultural milieu of everyday life is essential in understanding public perceptions of risk. Social actors do not respond to risk as disparate reflexive agents, mechanically weighing up the costs and benefits of decisions. More exactly, people act as 'situated agents' within collective surroundings and networks. By placing overriding emphasis on the rational dimensions of reflexivity, the risk society perspective disconnects social beings from embedded traditions and customs (Furlong and Cartmel 1997: 113; Lash, 1993). It would seem reasonable to speculate that individuals are capable of alternating between perspectives on risk, depending upon cultural context and social situation (Vera-Sanso, 2000: 126).[13] As Lupton and Tulloch (2002b: 326) comment, 'risk knowledges are the products of ways of seeing, rather than being fixed in their meaning.' Further, the weight of research suggests that attitudes towards risk will be structured by wider cultural factors such as public morality, societal norms, political ideology and welfare provision (Caplan, 2000a; Taylor-Gooby, 1999; Thompson, 1983). The very diversity of national and regional conditions intimates that Beck's reflexive and vigilant risk actor is likely to be something of a fiction, with reality being decidedly more contradictory and abstruse.

CONCLUSION

In conclusion, the match between Beck's theoretical project and empirical research into risk perception is somewhat scratchy. Computing with the risk society thesis, both cognitive and cultural studies suggest a relative rise in risk awareness in western cultures (Marris and Langford, 1996; Slovic, 1992). Furthermore, ethnographic studies have served to highlight the development of reflexivity through engagement with issues of risk (Macgill, 1989; Reilly, 1999). However, we have also stumbled upon notable contradictions and departures. Most appreciably, research indicates that risk perceptions are more fluid and culturally variable than the totalising narrative of the risk society grants. Needless to say, recognition that under-standings of risk are canalised by class, gender, age, ethnicity and geography would detract from the universal applicability of the risk society plot.

We have also thrown open a series of methodological problems which arise out of the dominant tradition within risk research. Historically, the method utilised within risk research has been remarkably homogeneous, with the majority of empirical studies working within the parameters of an objectivist-rational framework. As a result, the social and cultural underbelly of risk perception has been neglected. Cognitive-scientific studies of risk perception can tell us *how* risks are categorised, but they are less precise about exactly *why* these categorisations occur. Ideally, a comprehensive programme of ethnographic research is needed to ascertain how cultural conditions foster perceptions of risk and why understandings vary over time and place. The extent to which risk perceptions are structured by combined variables such as class, gender, ethnicity and age would seem to be a rich well of inquiry. To maintain credibility, future research needs to be attentive to the collaborative context in which individuals encounter risk. Simply extricating respondents from normalised routines and instructing them to respond to hypothetical questions is not a particularly profitable or thorough method of accessing cultural understandings of risk. As will be discussed in Chapter 7, both Beck's work and the psychometric tradition fail to adequately account for the collective and symbolic aspects of risk perception (Alexander 1996; Lupton, 1999a: 82). Risks are not approached in objective isolation by lay actors, but in situated settings with accumulated sets of cultural luggage. In addition to addressing the individual-rational dimension, we need to be sensitive to the melange of social, economic and cultural factors which underpin public perceptions of risk. As Langford et al. (1999: 33) note, public understandings of risk are composed of a 'collage of outlooks, predispositions, relationships, and structures all relating to each other in complex ways, like stars in a rotating galaxy'. Whilst this implies a less compact and messier idea of social reality than the risk society model, it may nonetheless provide a more accurate reflection of the complex nature of public perceptions of risk.

6
Living with Risk

Having built a bridge between the risk society thesis and empirical research into public perceptions of danger, we are now in a position to consider both the material impacts of risk and individualisation on routine cultural practices and the broader relationship between risk, trust and reflexivity. Allowing for the bulkiness of such an inquiry, this task will be stretched over the remaining three chapters. Here, we concentrate on the tangible effects of individualisation on everyday relationships within key social domains. In Chapter 7, we move from the material to the abstract, developing a thicker discussion of the interplay between manufactured risks, expert systems and public reflexivity. In the final chapter, we turn to the political dimensions of reflexivity, examining the extent to which the cyclic appearance of risk stimulates public engagement in emancipatory politics.

Despite the bold narrative of the risk society, the fundamental essence of a given culture – never mind an entire society – is difficult to trace. Our own selective interpretations are bound to feature and we cannot hope to do justice to the full spectrum of values, activities and practices. Thus, rather than striving to capture the generic nature of 'global risk experience', the more bounded undertaking of this chapter is to place the theory of risk society over the top of contemporary cultural trends. As a means of examining the extent to which Beck's argument reflects the ebbs and flows of modern life, we review the structural changes and residual continuities resulting out of the diffusion of risk and individualisation. Mirroring the con-figuration of the risk society thesis, we focus in on work, the family and relationships as barometers of institutional change. Although precedence is given to the transformative effects of the individuali-sation process, we also assess the extent to which changing patterns of risk distribution alter the dynamics of lived experience in contemporary society. To this end, we begin by conveying Beck's claims about the effects of individualization on social formations. This general discussion is followed by a more specific sketch of the impacts of risk and individualisation on class structure, interpersonal relationships and employment relations. Having followed the perimeter of the risk society argument a critical assessment will be undertaken, drawing on a mix of theoretical and empirical evidence.

THE INDIVIDUALISATION OF LIFESTYLES

As outlined in Chapter 1, the risk society is propelled by the diffusion of two intersecting processes. The first involves the negotiation of individualised life paths that are increasingly reliant on individual choice and reflexivity.[1] The second runs along the lines of risk distribution, indicating a gradual evolution of the distributional logic in society. In the risk society, these macro processes commingle, generating the emergence of a global culture of change and uncertainty. For Beck, individualisation serves to break up social structures and interrupts established customs (Beck, 1992: 127; 1997: 94; 1998: 169). As highlighted in earlier chapters, individualisation is a composite and multivalent process. The extensive scope of individualisation has led to the process being associated with a range of social fields, from personal relationships to political engagement (Heelas, 1996). Seen from the risk society prospect, individualisation describes the perpetual requirement to negotiate and select courses of action:

Individualization means that each person's biography is removed from given determinations and placed in his or her hands, open and dependent upon decisions. The proportion of life opportunities which are fundamentally closed to decision-making is decreasing and the proportion of the biography which is open and must be constructed personally is increasing. Individualization of life situations and processes thus means that biographies become *self-reflexive;* socially prescribed biography is transformed into biography that is self-produced and continues to be produced. (Beck, 1992: 135)

Beck is particularly keen to accentuate the degree of personal choice involved in the construction of life biographies. The comprehensive spread of individualisation means that personal decision making becomes an inescapable aspect of contemporary life: 'people are damned to individualization, using Sartre's terms' (Beck, 1998a: 33). As individuals become untied from the certainties of collective structures, everyday life becomes contingent on an infinite process of decision making. Central to the detraditionalisation of cultural experience is the questioning of traditional gender and occupational roles and an increased emphasis on identity choices. In line with the current sociological inclination, Beck believes that the localising ties of family, community, work and religion slackened in the second half of the twentieth century. In this respect, Beck's approach to changing patterns of identity formation is consistent with the work of Anthony Giddens (1990; 1991; 1994). Both Beck (1992; 1999) and Giddens (1994; 1999) maintain that individualisation involves the disembedding of social relations, in the geographical and the interpersonal sense. For Giddens (1991: 47), the bond between intimacy and propinquity has been broken; it is no longer possible to permanently connect self with place. In late modernity, people do not customarily work in their places of birth and families and friends no longer live within close proximity:

> In the global age, one's own life is no longer sedentary or tied to a particular place. It is a travelling life, both literally and metaphorically, a nomadic life, a life spent in cars, aeroplanes and trains, on the telephone or internet, supported by the mass media, a transnational life stretching across frontiers. (Beck, 2002: 25)

As a result of personal mobility and the stretching of social networks the cohesiveness of the socialisation process is endangered. Previously secure sites of solidarity recede, support networks dissolve and individuals are encouraged to turn inwards toward personal decisions and self-resources (Beck, 1992: 92). Although the open-ended nature of identity construction presents individuals with greater scope for creativity, it is also productive of unsettling dilemmas: 'all too swiftly the "elective", "reflexive" or "do-it-yourself" biography can become the breakdown biography' (Beck, 1999: 12). As Beck sees it, people are handed – and subsequently obliged to manoeuvre – a double-edged sword. One blade cuts greater choice and autonomy, the other carries the burden of continual decision and responsibility.

THE UNBINDING OF SOCIAL STRUCTURES: FAMILY, CLASS AND WORK

Central to the risk society narrative is the notion that risk and individualisation are fundamental levers of cultural change. The economic and techno-scientific orientation of the capitalist system ensures that manufactured dangers seep through institutions and everyday practices. At the same time, the process of individualisation cascades over a series of social tiers, reinforcing the need for personal planning and decision making. In order to grasp the impressions made by individualisation and risk, Beck examines structural changes in the family, social class and employment practices.

In both *The Normal Chaos of Love* (Beck and Beck-Gernsheim, 1996) and *Individualization* (Beck and Beck-Gernsheim, 2002) the family and interpersonal relationships are placed under the microscope. As opposed to the collective experience promoted by social class or the nuclear family, individualisation promotes the self-management of lifestyles. Beck (1992: 9) believes that a combination of shifts – for example, in the standard of living, educational opportunity and geographical mobility – have dissipated class distinctions, leading to a 'diversification of lifestyles'. These alleged transformations in social structure are driven by the individualisation process, which challenges taken-for-granted assumptions and stimulates a new mode of socialisation:

> What has manifested itself over the past two decades in Germany and perhaps in other industrial states as well can no longer be understood within the framework of existing conceptualizations. Instead, it must be conceived of as the beginning of a new mode of societalization, a kind of 'metamorphosis' or 'categorical shift' in relation to the individual and society. (Beck, 1992: 127)

At a structural level, the unbinding of social structure yields both fragmentation and cultural diversity. In industrial society, class and the nuclear family acted as linchpins of social organisation and filters of occupational destination. In the risk society, the linchpins rust away and social opportunities become individualised (Beck and Beck-Gernsheim, 2002: 4). Beck argues that, in contemporary western cultures, the permanence of marriage bonds cannot be assumed or expected. Instead, building and maintaining 'elective' relationships becomes a compulsory activity. Far from accepting matrimony determined by geography, class and parental aspirations, people opt for partnerships that do not inevitably result in wedlock. In a similar vein, where once class could be assumed to be a mechanism of social association and integration, in contemporary society, class structure becomes more labile and class identities more indistinct. In effect, the individual has replaced social class or the nuclear family as the fundamental unit of social reproduction (Beck and Beck-Gernsheim, 2002).

In addition to class and familial structure, Beck views the labour market as a key site of individualisation and a motor of risk distribution (Beck, 1992; 1998; 2000a). In *Brave New World of Work* (2000a), transitions in employment relations are tied to the historical scaffold of the risk society thesis. According to Beck, changes in working practices, class structure and the family are indicative of a move away from a 'Fordist regime' and towards a 'risk regime' (Beck, 2000a: 67). The Fordist regime – common to western cultures from the 1950s to the 1970s – is characterised by full employment, rising living standards and job security. During the Fordist era, employee rights were relatively strong, bolstered by trade union activity, collective bargaining and Keynesian macro-economic policies. Beck believes that the institutional management of the system ran comparatively smoothly, with Fordist methods functioning as an efficient distributor of 'social goods'. In the boom period of mass production western citizens bought in to the capitalist system, at an ideological and a material level. During the Fordist era the majority supported the principles underpinning mass production and possessed a general faith in progress (Beck, 1998a: 42). Trust relations between individuals and the state were strong and a cultural consensus developed around the ability of technology and industrialism to provide a high standard of living for all. For Beck, the cementing of universal social interests in industrial society was demonstrated by relative harmony in labour relations and the acceptance of standardised employment contracts, rights and obligations:

> The Fordist regime ... rests upon the fact that the principle of mass standardization applies to both production and consumption. Labour and production are geared to large model runs of cars, refrigerators, washing machines and the like, which allow rapid increases in productivity and profits and, via rising wages, also in mass consumption. This form of production, work and consumption created a society in which people's lives were as highly standardized as the sheet metal from which the cars were welded together. (Beck, 2000a: 68)

Although the institutional structures of the Fordist regime were transiently successful, the collapse of the system in the 1970s was signalled by the widespread dispersal of 'social bads', such as rising unemployment, redundancies and the disintegration of job security. In the 1980s, the consolidation of a highly competitive global market led to changes in organisational methods, such as the utilisation of computer technologies and automated production methods.[2] Beck avers that the infusion of new technologies into the workplace has been detrimental to employment prospects. Aligning himself with André Gorz (1982; 1988; 2000), Beck (1998: 58; 2000a: 5) contends that the use of automated machinery has generated deskilling and led to substantial reductions in labour requirements. Beneath the shiny veneer of technological flexibility lies the risk of unemployment:

> Here we have the new law of productivity that global capitalism in the information age has discovered: fewer and fewer well-trained and globally interchangeable people can generate more and more output and services. Thus, economic growth no longer reduces unemployment but actually requires a reduction in the number of jobs. (Beck, 1998: 58)

Insofar as jams in the Fordist machine were predictable and rectifiable, the risk regime produces volatile and unanticipated effects. The global interconnection of capitalist markets means that product demand, employment requirements, rates of exchange and stocks and shares all become precarious entities. Concomitantly, the world of work becomes less stable, skills and labour are rapidly rendered obsolete and welfare cover contracts. Where the Fordist regime is open to national governance and intervention, under the risk regime 'nothing can be foreseen or controlled' (Beck 2000a: 77). As the risk regime extends, employers and the nation state lose structural autonomy and control. Agentially, traditional networks of security and collectivism collapse and individuals are forced to assume responsibility for mapping their own life biographies:

> For a majority of people, even in the apparently prosperous middle layers, their basic existence and life-world will be marked by endemic insecurity. More and more individuals are encouraged to perform as 'Me and Co', selling themselves on the market place. (Beck 2000a: 3)

A SHIFT IN EVERYDAY LOGIC: FROM GOODS TO BADS

As discussed in Chapter 1, Beck believes that economic and techno-scientific modernisation has produced injurious effects within various social fields. The side effects of capitalist development reverberate through everyday domains, such as the workplace, family and community. However, Beck is not simply arguing that the contemporary world is a dangerous environment to live in. At a deeper level, the fecundity of risk fuels a sea change in public perceptions and motivations. To concretise this shift, the risk society narrative outlines a fundamental movement away from the 'logic of goods' and toward the 'logic of bads' (Beck, 1995a: 78). As described in the opening chapter, the two logics are linked up to particular stages of modernity and related to changes in the composition and consequences of risk. To distinguish between different epochs, Beck refers to a triad of defining features. In simple industrial society, a lack of social goods leads to feelings of hunger, which drive political concerns about scarcity. By contrast, in the risk society, an excess of social bads leads to feelings of anxiety, which fuel concerns about safety (Beck, 1992: 49). In classic industrial society, socio-political objectives are geared around the way in which the 'cake' is divided up. That is, people are engaged in acquiring sufficient pieces of pie to have a happy and fulfilling life. However, in the risk society, the cake becomes poisoned, radically altering the purpose and pattern of distribution. As a consequence, people become less concerned about acquiring 'social goods' and more concerned with avoiding 'social bads' (Beck, 1992: 20). The rising tide of cultural anxiety about risk not only signifies that institutions are failing in their role as guarantors of public safety. In many instances, the general public opine that institutional interventions have actually exacerbated existing problems (Beck, 1995a: 122). As recounted in Chapter 2, institutional action and intransigence produce knock-on consequences, leading to the creation of vicious circles of environmental risk. For instance, a failure to combat rising pollution levels leads to a larger hole in the ozone layer, which leads to higher incidence of skin cancer. This in turn generates unmanageable burdens on national health systems, infinite waiting lists, perfunctory treatment and a poorer quality of public health. Due to the escalation of manufactured risks, the balance of public and political concern shifts from a positive logic of goods acquisition to a negative logic of bads avoidance. As we saw in Chapter 1, this transition reconfigures social understandings of safety. Instead of the sectoral patterns of security common to class society, the dangers of the risk society are universal. Given the global reach of manufactured risks anyone and everyone can be exposed (Beck, 1992: 36). Affluent groups and regions previously sheltered from the hazards of industrial society, become subjected to 'boomerang effects':

> Formerly latent side effects strike back even at the centres of their production. The agents of modernisation themselves are emphatically caught in the maelstrom of hazards that they unleash and profit from ... earth has become an ejector seat that no longer recognises any distinctions between rich and poor, black and white, north and south or east and west. (Beck, 1992: 37)

As we shall see in Chapter 8, the unavoidable conflicts and controversies which spring up around social bads make the risk society an inherently political epoch (Beck, 2000c: 220). The generalised distribution of risk leads to heightened public consciousness and a more reflexive culture (Beck, 1994: 6). Via the knowledge accumulated through habitual negotiation of risk, people come to question the validity of dominant institutions. As modernisation evaporates the certainties of industrial society – lifetime careers, the nuclear family and class identities – the very elements of everyday life become issues of dispute and contestation (Goldblatt, 1995: 163).

INDIVIDUALISATION AND EVERYDAY LIFE: THE CASE FOR

By way of exegesis, we have relayed the material effects of individualisation on social structures and outlined the consequences of generalised patterns of risk for the balance of public concerns. In the remainder of the chapter, we cast a more discerning eye over Beck's argument, focussing on the impressions made on everyday life by risk and individualisation. To facilitate balance, we consider both supporting and opposing evidence of transitions in employment, class and interpersonal relationships. First, in accordance with the risk society thesis, we present the evidence of an increasingly individualised and risk-soaked culture. Second, we consider viewpoints which question the breadth and depth of the impress made by individualisation. Third, we contest the notion of a fundamental shift in the distributional logic of risk, drawing on existing critiques.

Without doubt, Beck and his partner Beck-Gernsheim have made considerable strides in encapsulating the changing dynamics of family life and personal relationships. In contemporary western society, the family no longer resembles the nuclear ideal of the Fordist era. As a package, the sole male breadwinner, gendered child-rearing responsibilities, clear divides between domestic and formal work and rigid age hierarchies have become the exception rather than the rule. Retrospectively, it is true to say that, in the 1950s and 1960s, families were relatively cohesive, localised and formally based on the model of the nuclear family. Since then, a number of factors have chipped away at this 'ideal' in Europe and North America including social mobility, the quest for individual fulfilment, the normalisation of divorce, the loosening of class bonds and the spacing out of work. As a result, gender roles, the family and personal relationships have all undergone reformation and reordering. It is now widely accepted that loving relationships require regular maintenance and that conditions of intimacy may be renegotiated between one relationship and the next. Thus, in contemporary European cultures, greater weight is indeed placed on establishing 'elective affinities' (Beck and Beck-Gernsheim, 2002: 85). The accentuation of choice and preservation has generated different attitudes towards biological reproduction and parenthood. In western cultures, many couples are reversing the baby-boomer trend and opting to have children later in life. By 2001, British women aged between 30 and 34 were statistically more likely to give birth than those aged between 20 and 24 (Social Trends, 2001). Furthermore, the trend toward raising fewer children has impacted on the composition of family structure. In Britain, the average household size has downsized from 4.6 to 2.4 in the last 30 years. Moreover, for every two couples marrying in 1999 one was registering a divorce, again indicating greater fluidity between relationships (Social Trends 2001). There has also been a notable increase in lone-parent families and a rise in the number of reconstituted families (Sherratt and Hughes, 2000: 57). Recently, gay rights campaigners have finally succeeded in gaining legal recognition for same-sex marriages. These trends and figures do indicate that personal relationships are characterised by higher doses of choice, change and decision than experienced in previous eras. Travelling along with Beck, it is probable that, beside choice and opportunity, the changing structure of interpersonal relationships has also nurtured insecurities. Measured against previous epochs, individuals in western cultures do have greater freedom to construct identities and to select loving relationships. However, the very act of choosing invokes the lumber of personal responsibility: 'nothing is immutable, there are no permanent alliances and no eternal verities; more than ever before the future appears to be riven with uncertainty' (Wilkinson, 2001: 29).

As well as impacting on the form of personal relationships there is also evidence to suggest that work has become a distributive axis of individualisation. On balance, it is tempting to concur that automation and computerisation have eliminated more jobs than have subsequently been created. Nevertheless, it seems reasonable to assume that a constellation of factors have heightened the risk of unemployment in western Europe. The rise of the so-called East Asian 'tiger economies', financial mismanagement, union weaknesses and state intransigence have all played an important role in bearing current job insecurities in the West (Hutton, 1996: 257–85). As will be illustrated, Beck's alacrity to knit employment patterns to risk disregards the diverse factors which have created current labour-market uncertainties. For Beck, discussions of risk cannot be divorced from the powerful and uprooting force of individualisation. For many people, the insecurity associated with work has engendered a series of risks and dilemmas. Responding to the insecurities associated with employment contracts, wages and child-care has become the 'stuff of life' at the turn of the millennium. In contemporary western society, the unemployed demand the right to work, whilst many of the employed keep an anxious eye out for the twist of fate which may familiarise them with the peculiarities of the welfare system. Despite the lack of precision, opinion polls in the UK report that up to half of the workforce are either 'very' or 'fairly' concerned about the threat of unemployed (see Doogan, 2001: 420).

The introduction of new workplace technology has doubtless complexified the division of labour and opened up the possibility of employment choice. In pre-industrial cultures trades were few in number, compared to the endless band of occupations available today. There is also evidence to suggest that changes in employment contracts have individualised employment costs and benefits (Furlong and Cartmel, 1998). The utilisation of self-employed, temporary and part-time staff has allowed various costs to be transferred to individual workers. As Beck points out, sick pay, training and pension provisions have become the responsibility of the temporally flexible employee, rather than the employer or the state (Beck, 2000: 53–4; Dawson, 2000).[3] The changing quality and shape of workplace relations have also affected the cohesiveness of class structure. Prior to the 1980s, European labour markets were characterised by a relatively strong demand for unqualified school leavers working in large industrial units. Since the 1990s patterns of labour demand have changed significantly, with opportunities for young workers being located in numerically smaller workplaces. The demand for flexible specialisation and the increased use of part-time and temporary employment contracts have weakened collective employment experiences (Furlong and Cartmel, 1997: 3). Furthermore, changes in the nature and technologies of labour have left open scars in industrial heartlands specialising in manually produced goods, such as coal, steel or ship building. In these communities, the material cushions provided by the communality of class have been scattered. In Britain, the decimation of towns and cities built around single industries has cut off the support mechanisms traditionally nurtured within working-class cultures. Trade-union activities, community groups and social clubs are receding features within many former industrial areas. Concurring with Beck, it is likely that cultural fragmentation resulting from changes in the labour process has yielded a relative decline in class identity, particularly amongst young people. Youth unemployment has risen steeply in most European countries and training schemes, further education, and a desire to travel and work abroad have further dislocated the pathways of collective mass experience. The theory of individualisation appears then to be reflective of dominant cultural trends. In early twenty-first-century western cultures, social class has declined as a visible facet of self-identity, the nuclear family is no longer a fundamental structural given and career change and geographical mobility have become accepted features of working life. Thus, it is reasonable to suggest that a symbiotic relationship exists between individualisation and social change. As energies are concentrated on personal education, job maintenance and relationship building, people are becoming psychologically conditioned to individualisation and are driving the process forward through social motions (Furlong and Cartmel, 1997; Wilkinson, 2001: 30). However, as we shall see, whilst evidence can be mobilised to endorse Beck's position, several knotty issues arise out of the usage and application of individualisation in the risk society thesis.

INDIVIDUALISATION AND EVERYDAY LIFE: THE CASE AGAINST

On the basis of our discussion, it would appear that individualisation – understood as the intensification of individual decision making and the concomitant dissolution of traditional structures – is gathering momentum in western cultures. Having assessed recent transitions in employment relations, the family and personal relationships, it has been demonstrated that the process of individualisation is generating changing cultural practices and personal expectations. Nonetheless, a number of areas of concern surface around the thick delineation of individualisation and the social depth of the process.

First, it is worth pointing out that individualisation, like risk, functions as a multipurpose instrument. It is only in piecing together the risk society thesis that the full ambit of individualisation is revealed. Individualisation is constituted by a rise in lifestyle choices; the fragmentation of cultural experience; a proliferation of social risks; greater personal responsibility and accountability; the undermining of class identities; social disembedding and the development of diverse and reflexive life paths. Although such definitional diversity may reflect the polymorphous nature of current cultural transitions, it also raises questions of sociological validity. It would be difficult to falsify the diffusion of the individualisation process, yet, given the extensive boundaries of definition, it is hardly surprising that supportive evidence can be marshalled. However, by the same token, individualisation is equally difficult to 'prove'. Taking on board the breadth of Beck's understanding, we might reasonably expect to find a ready supply of supportive examples.

The rather imprecise meaning of individualisation in the risk society thesis leads us to the conundrum of how best to calibrate the depth and the effects of individualisation in contemporary culture. Although Beck does attempt to deconstruct the elements of individualisation, the lived experience of individualisation is tacitly assumed, rather than substantiated by empirical analysis. Thus, the issue of how socially meaningful individualised experiences are *in relation* to collective experiences remains unresolved. Because Beck is predisposed towards evidence of individualisation, the risk society thesis presents a lopsided account of cultural experience which highlights uncertainty and hides away stability.

Even if we are to turn a blind eye to the selective deployment of evidence in the risk society thesis, an important question mark still lingers over the structural significance of the process. Insofar as Beck informs us that individualisation radically alters the structure of society, it must be remembered that the development of modernity has produced various forms of social differentiation (Polanyi, 1975). Indeed, the diversification of life trajectories and the decline of tradition have been long-standing social concerns, expressed in the classical sociology of Simmel, Durkheim and Weber. This brings to the surface issues of relativity, particularly as regards the *speciality* of individualisation. Are we currently witnessing fundamental social reconfiguration, or are contemporary forms of individualisation simply the continuation of a process which is centuries long? Beck tends to present individualisation – and risk generally – as 'new' cultural experiences, which fundamentally alter human relationships. However, it is worth reminding ourselves that both individualisation and risk have a decidedly long history and are organic processes.

A further blemish in the risk society thesis stems from Beck's habit of utilising hand-picked illustrations to support a general argument. Reflecting on the social circumstances unique to the risk society, one can propound a range of counter examples which question the haecceity of contemporary patterns of individualisation. In assessing the extent of individualisation, we need to be receptive to social order as well as change. Adopting tunnel vision, Beck neglects the cohesion of social structures and flattens the complexities of social reproduction. By dissembling structural continuities and exaggerating the novel features of individualisation, Beck falls into overblown claim making: 'The place of traditional ties and social forms (social class, nuclear family) is taken by secondary agencies and institutions, which stamp the biography of the individual and make that person dependent upon fashions, social policy, economic cycles and markets' (Beck, 1992: 131).

Even allowing for hyperbolic style, such generalised claims serve to gloss over evident continuities in patterns of social reproduction in western cultures. To argue that class and the nuclear family are losing relative cohesion as agents of socialisation is one thing. It is quite another to suggest that these structures are being replaced by 'secondary agencies'. In a time of extensive social change, it must be recognised that long-standing class divisions endure. In Britain, the wealthiest one per cent owns 23 per cent of total marketable wealth, with the poorest 50 per cent owning less than 6 per cent.[4] Similarly, although the family has been a site of flux, empirical evidence also indicates marked continuities in both form and cultural significance. A recent *Social Trends* survey (1999: 43) reports that 79 per cent of British families are couple families, living with or without children. In addition, the ideological role of the family in influencing and ordering social life has been sustained (McGlone et al., 1996; Sherratt and Hughes, 2000). Contrary to the risk society perspective, research indicates that young people are actually remaining dependent on their families for extended periods (Jones, 1995). Whilst the expansion of higher education has meant that young people leave the parental home at an earlier age than their forebears, one in three will subsequently return (Furlong and Cartmel, 1997: 45).[5] Looking towards the future, the growing trend of 'putting back' marriage, coupled with astronomical rises in house prices implies that young people may be destined to spend longer periods of time/life in the family environment.

The reproduction of continuities in class and familial structure highlights the problems of generality that bedevil the risk society thesis. These problems are particularly pronounced in the construction of regimes of employment. Although Beck's 'risk regime' speaks to the uncertain experiences which filter through the labour market, the historical distinctions drawn are heavy-handed. Rather than a 'radical rupture' occurring in the sphere of work, it is more likely that current instabilities are part of an ongoing pattern of reorganisation. Further, the caricatured features of the risk regime are too broad to resemble a scattered and culturally differentiated world of work. As Will Hutton's (1996) 'forty, thirty, thirty' model suggests, socio-economic groups will encounter different employment and life experiences. Regardless of whether class identity has waned, the degree and intensity of individualisation will be mediated by existing inequalities of class, gender, ethnicity and age. In addition, internal working practices, such as quality of engagement, working conditions and relations with colleagues will influence the lived experience of individualisation (Emslie et al., 2000). For his part, Beck accentuates the individualising aspects of employment and disregards the collectivising dimensions of either paid or informal working practices; in the risk society everyone seems destined to share a universal individualised experience. In casting individualisation as an evenly falling snow, Beck rather flattens cultural, economic and regional differences.[6] In his desire to script individualisation as a universal phenomenon, Beck elides the fact that the degree of exposure to the process is governed by geographical location. Although Beck makes reference to the 'global' spread of the individualisation process, he rarely ventures beyond cultural changes in his native Germany (Beck, 2000a: 145; Marshall, 1999: 155). Of course, the degree to which employment practices act as a catalyst for individualisation can only be properly evaluated with reference to appropriate geographical, social and economic contexts.

Similarly, whilst labour markets in western Europe might well have become uncertain domains, it is unlikely that the diversification of risk has equalised employment experiences. Despite the occasional appearance of boomerang effects, uncertainty and risk will gravitate towards the 'bottom' 30 per cent of unemployed, part-time and temporary workers. There is little evidence to suggest that employment-based inequities have been evened up. Those from working-class backgrounds still feel the effects of job insecurity most acutely (Day, 2000; Furlong and Cartmel, 1997: 27–40; Perrons, 2000). In 1992 – the year the English translation of *Risk Society* appeared – just 2 per cent of school leavers with parents from professional occupations were unemployed by the following spring. For those with parents in manual occupations, over 10 per cent were jobless (Courtney and McAleese, 1993). As a generic group, manual workers are more likely than non-manual workers to experience long tracts of unemployment (Wilkinson, 2001: 55). In times of economic recession, the least educationally qualified will be the most susceptible to unemployment (Gangl, 2001: 67). Cognate layers of employment stratification also follow the fissures of ethnicity and gender. For instance, the British unemployment rate for white males is 9 per cent, for Black-African men it is around 28 per cent (Denscombe, 1998: 14). Meanwhile, there are double the number of women than men in temporary forms of employment, with women being five times more likely than men to be engaged in part-time work (Wilkinson, 2001: 73). Overall, the female wage in Britain is just over three-quarters of the male average (Denscombe, 1998: 12). Thus, employment paths in Britain still appear to be strongly determined by class, ethnicity and gender. Whilst the sphere of employment can be depicted as a site of individualised experience, actual life chances are wedged within the grooves of traditional inequalities:

> Although changing school to work transitions have led to an increased risk of marginalisation, risks continue to be distributed in a way which reflects social divisions characteristic of the traditional order. In other words, it is still possible to predict labour market outcomes fairly accurately on the basis of social class (via educational performance) and gender. Indeed, while the breakdown of collectivist traditions created the illusion of individuality, these changes have had little effect on processes of social reproduction. (Furlong and Cartmel, 1997: 109)

As Furlong and Cartmel suggest, fluctuations in the labour market have not radically altered patterns of risk distribution. Disadvantaged classes still tend to experience higher levels of unemployment, are more likely to constitute peripheral workforces, and habitually live with job insecurity. The class-related dispersal of employment risk pulls up the roots of the risk society thesis. If the universalisation of 'employment risk' is a perceptual rather than a concrete category – that is, if the experience of the risk society is more about *feeling* job insecurity than being rendered redundant – the purchase of the argument is diminished. In truth, Beck's take on 'employment risk' is an admixture of contemporary review, historical contrast and future prediction. Unfortunately, this distinctly pied approach disguises different levels of risk and conflates perception and actuality. In some cases, fears of unemployment will not be congruous with the probability of job loss. As Doogan (2001: 436) reasons: 'even when people are confident about keeping their job, they can still worry about the prospect of losing it'. Bluntly put, there is clear blue sky between risk as an unsettling possibility and being handed the P45. Thus, the risk society thesis imperfectly translates the specific into the general. Ultimately, Beck's portrayal of employment risk is built around on the experiences of particular workers, in particular sectors:

> Attention is focused ... on one or two sectors that have large representations of young workers that symbolise the McDonaldisation of service sector employment ... to generalise from this and establish some post-Fordist labour market regime is to provide a distorted perception of reality. (Doogan, 2001: 434)

REVISITING THE LOGIC OF DISTRIBUTION

The universality Beck attributes to the individualisation process is also implanted in the theory of distributional logic. In this section we dispute the evidence of a universal shift in distributional logic by pursuing four interrelated avenues of critique. First, the movement from industrial to risk society rests upon the shaky premise that social bads have become universal and unavoidable facets of everyday existence. Although the equalisation of risk distribution appeals to egalitarian principles, the uneven layering of social bads rebuts Beck's argument (McMylor, 1996). In contemporary society, economic factors still govern personal choices and the range of risk reduction strategies available, indicating that risk tracks poverty and inequality (Culpitt, 1999: 21; Draper, 1993). On terra firma, the boomerang is rarely cast out in such a way that it returns to the pitcher. As Scott points out, economically dominant groups previously insured against poverty remain able to buy their way out of risk situations: 'The wealthy were protected from scarcity and remain protected from risk; "protection" here being understood as "relative protection". Smog is just as hierarchical as poverty so long as some places are less smoggy than others' (Scott, 2000: 36).

The unrelenting reproduction of inequalities in western Europe suggests that social class remains a significant yardstick of life chances in contemporary society. Access to material resources continues to be the key determinant of action. It is therefore improbable that the radical restructuring of cultural experience outlined in the risk society thesis has transpired (Elliott, 2002: 304). Taking this argument a step further, we can justifiably question the exceptionality of the risks and uncertainties related in the risk society narrative:

> Much of what Beck describes has long been standard for those without much money or control over their lives. Many, perhaps most, individuals have traditionally found it difficult to read the future, to remain in one place with their families and friends; in brief, to determine their own lives. (Day, 2000: 51)

Second, several theorists have pointed out that the theory of distributional logic makes no reference to empirical evidence and simply assumes that public perceptions of risk have been fundamentally made over (Draper, 1993; Hajer and Kesselring, 1999: 3). Disregarding the process of validation, Beck maintains that risk has worked its way to the forefront of individual consciousness. However, this higher-order cognitive positioning is theoretically imputed without recourse to evidence: 'fragments from the empirical world intrude only as illustration or example' (Leiss, 2000: 7). Whilst it is laudable insight that western cultures have witnessed an increase in public awareness of manufactured risks, there is scant evidence to suggest that this has come at the expense of concerns about the inadequate distribution of goods. Empirical research does point towards an overall rise in environmental awareness in western cultures, but it cannot be consequentially inferred that this has taken up the cognitive space previously held by issues of material inequality. As revealed in Chapter 5, the risks people *fear* the most are not necessarily those they *focus* on the most amidst the trials and tribulations of day-to-day life.

Third, rather than engaging in systematic academic investigation, Beck falls back on the trusted icons of destruction as emblems of social concern. Unfortunately, this strategy results in the qualitative distinction between goods and bads being rendered dependent on a limited selection of well worn examples. Widespread public concern about global risks cannot be established by simply reciting a hymn-sheet of worst imaginable accidents. Even within this thin band of exemplars, there are sizeable differences in the scale of danger and the magnitude of consequences. By collapsing hazards into a standardised category of social bads, disparate risks are injudiciously compacted (Scott, 2000: 36).

Fourth, as hinted at in Chapter 2, methodological problems arise out of the distinction between class and risk. Although the use of binary logics enables theoretical contrast, the separation between modes of perception exaggerates difference and masks historical similarities:

> 'I am afraid', for Beck the motto of the risk society, is no less appropriate to class societies even if the focal point of anxiety has shifted. Fear of hunger, like the risk of ecological catastrophe, is most of the time probabilistic. (Scott, 2000: 36)

Collectively, these four strands of critique bear testament to a hiatus between the lofty theory of the risk society and the assorted cultural practices which constitute everyday life. There is scant empirical evidence to support Beck's claim of a radical shift in social logics in western cultures. It would seem more credible to argue that traditional issues of poverty and inequality intermingle with concerns about manufactured risk, both at a political and a personal level.

CONCLUSION

In assessing the cultural inroads made by risk and individualisation we have journeyed through a series of social fields. In concordance with the risk society thesis, it has been noted that individualisation is becoming a common facet of everyday life, propelled by global-isation, the labour market and the changing dynamics of personal relationships. In this respect, the risk society perspective is attuned to the constant demands of planning and shaping the future placed on the individual in contemporary society (Lupton, 1999b: 67). Furthermore, Beck's appreciation of the inherently unstable quality of 'tightrope biographies' (Beck and Beck-Gernsheim, 2002: 2) keys in with the uncertainties and insecurities faced by many people in western cultures. By way of critique, it has been noted that Beck attributes sweeping social power to individualisation without reference to empirical research (Engel and Strasser, 1998: 97). Indeed, the risk society perspective tends to resort to theoretical exaggeration in order to compensate for empirical deficiencies. In nurturing novel sociological ideas, there is a need to remain sensitive to what is actually going on, not what makes for a tidy theoretical argument. While there is little dispute that individualisation has become an integral component of the socialisation process, Beck's broader claims about the power of individualisation need to be reined back. Traditional social structures such as the family, work and class are certainly in flux, but it is unlikely that they are being dissolved by a master process of individualisation. The non-specific usage of individualisation obfuscates the degrees of advancement made by various dimensions of the process and the patchiness of cultural coverage. It is expectable that different aspects of individualisation will sit more readily on some groups than others. Thus, a more systematic empirical approach is needed to establish both the *extent* and the *effects* of the individualisation process. In particular, greater attention must be paid to entrenched forms of social stratification and the contextual 'lived' dimension of individualised experience. Here, it has been argued that embedded layers of stratification and cultural identities will condition encounters with individualisation. Furthermore, the experience of individualisation will be affected by personal characteristics and preferences, self-resourcefulness and the robustness of local networks.

Turning to the marks made by risk on everyday activities, the picture is again mixed rather than uniform. In support of Beck's argument, recent surveys and studies have recorded high levels of public awareness about the detrimental consequences of environmental risks (see Macnaghtan, 2003). Furthermore, in times of heightened alert in the West about future terrorist attacks, levels of public risk consciousness are arguably as developed today as at any other point in history. However, there is analytical value in differentiating between risk as a perceptual feature and risk as an ultimate outcome. As will be elaborated in Chapter 7, Beck is too quick to conflate probability with impact. It must be remembered that we inhabit something of an 'over-anticipated world' in which the facility of institutions to identify risk may cause fear and disquiet (Woollacott, 1997). In times of high anxiety, we might comfort ourselves with the knowledge that both quality of health and longevity have increased markedly in the West in the last century. In the final analysis, the risk society thesis looks over pre-modern cultures with rose-tinted spectacles, concealing the fact that death, disease and hunger were widespread: 'if people managed to survive the high levels of infant mortality, then they most likely faced a short lifetime of constant physical suffering and fearful uncertainty' (Wilkinson, 2001: 34). In conclusion, Beck's claim that there has been a palpable movement in the social distribution of risk has justifiably raised the hackles of many critics. Contra the risk society thesis, there is scant evidence to suggest a shift from a differential class-based logic to a universal-ising logic of risk. Again, the general penumbra of global risks is mistaken for a tangible shift in distributional activity. Lamentably, Beck's theory of distributional logic does not allow for sufficient dif-ferentiation in both the manifestation *and* the experience of risk.

7
Risk, Reflexivity and Trust

Having investigated the material distribution of individualisation and risk, in this chapter we turn to the more abstract, philosophical issues surrounding the relationship between risk, trust and reflexivity. Building on the base constructed in Chapters 5 and 6, we will consider the extent to which risk influences the formation of self-reflexivity and impinges upon wider trust relations between experts and the public. After reciting the main features of Beck's approach, we evaluate the depth of public concern about risk, attempting to resolve the apparently paradoxical rise in insecurity during a phase of unrivalled safety. Following this contextual discussion we turn to the deliberative aspects of the hermeneutic process, elucidating attitudinal approaches suppressed by the risk society thesis. Having detected cultural variability in coping strategies, we make visible the aesthetic facets of reflexivity. Towards the end of the chapter, these two tributary critiques merge to form a broader discussion of the evolving trust relationship between institutional experts and the general public. Throughout the chapter, a collection of political issues will be turned up around the conflict between expert systems and lay actors in contemporary society. The wider ramifications of these issues are carried over into the final chapter in which we appraise the explicitly political angles of the risk society thesis.

INDIVIDUALISATION, UNCERTAINTY AND REFLEXIVITY

Up to now, we have prioritised the material dynamics of the risk society, analysing the transformative effects of risk distribution and individualisation on the family, work, class and welfare. However, in the risk society storyline, the intensity of individualisation and changing patterns of risk dispersal also produce marked changes in cultural attitudes, social knowledge and institutional trust. As we have seen, one of the most conspicuous effects of individualisation is that life biographies become self-determined (Beck and Beck-Gernsheim, 2002; Beck et al., 1994). Simply keeping oneself together in a changing cultural climate requires a continuous programme of reflexive monitoring. The emphasis on personal management promoted by the individualisation process is accentuated by the frequency with which risky decisions emerge in everyday life. In essence, the instability generated by individualisation is reinforced by the generalisation of risk distribution. As social bads appear across everyday cultural domains, individuals are required to construct strategies of self-defence to ward off risk. If we moderate alcohol consumption, we reduce the risk of liver failure. If we opt not to smoke, the risk of contracting lung cancer decreases. If we exercise regularly we reduce the possibility of suffering a heart attack. These everyday risk negotiations are supplemented by the daunting possibility of environmental and nuclear threats. The unremitting appearance of hazards demands that people engage in everyday risk assessments; reflexivity is 'built-in' to contemporary risks (Beck, 1992: 165). A combination of enhanced scientific knowledge about risk, political pressure by new social movements and media spotlighting propels the process of reflexive monitoring forward. Accordingly, a U-turn occurs in terms of people's aspirations and motivations. Where once individuals were driven by the positive goal of goods acquisition, the negative logic of bads avoidance rules thought and action. In the risk society, the growth and mobility of risk knowledge breeds both public uncertainty and reflexivity.

As far as uncertainty is concerned, the latent composition of risk means that 'manufactured uncertainties' are always potential 'dangers of the future'. Nobody knows precisely where or when risks will impact (Beck, 1999: 12). This projective variable has vital implications for both the formulation of specific understandings of risk and broader cultural preoccupations. Not only are risks difficult to codify, the frequency of uncertain incidents generates widespread cultural anxieties. Therefore, the omnipresent profile of risk produces a cumulative effect, resulting in a generalised climate of indeterminacy. As potential institutional routes to public safety are closed down, a culture of insecurity flourishes. The cloak of anxiety which hangs over the risk society, leaves individuals in a state of permanent watchfulness. In short, our minds become 'factories of fear'.[1] On the reflexivity side of the equation, as society moves away from traditional ways of understanding risk, rational and reflexive forms of thought increasingly structure social experience. Negotiating hazards becomes a customary feature of everyday life, with basic activities such as drinking water, travelling, eating food and sitting in the sun invoking impromptu risk assessments. Whereas traditional societies attributed risk to fate or religion, western cultures perceive human beings to be the drivers of destiny. Needless to say, this movement in social perspective has knock-on effects for the politics of responsibility. Rather than attributing hazards to nature and seeking refuge in fate, the routine appearance of risk necessitates strategies of self-management.

The risk society thesis is concerned not only with the individualisation and routinisation of risk, but also the broader transference of responsibility from institutions to individuals. Beck believes that the rolling back of the welfare state in capitalist cultures has led to a tipping of the scales of accountability. As governments divest themselves of responsibility for risk – through privatisation, promotion of private insurance systems and the withdrawal of state pensions – health risks are converted into baggage to be handled by the individual. For Beck (1987: 156), the migration of responsibility for risk has wide-ranging impacts on public attitudes towards politics, science and government. Public trust in social institutions is being stripped away as the frequency of risk incidents gnaws away at the legitimacy of expert systems. Beck argues that bungled risk management of regulatory agencies has served to generate public critique. In this respect, public reflexivity emerges as a response to both the intensity of risk incidents and ineffectual institutional responses. Due to the complexity and density of manufactured risks, regulatory agencies are prone to disseminating erroneous or contradictory information. For example, in Europe, public distrust in expert systems has been dented by a series of high-profile cases of institutional mismanagement, including Chernobyl and BSE. As a result, lay publics have become increasingly sceptical and combative towards institutional claims and promises concerning the economy, science, medicine and public health. Insofar as public trust in social institutions acted as a relatively stable given in industrial society, in the risk society continual media exposure undermines the power of expert knowledge.

SCIENTIFIC DEVELOPMENT, RISK AND FUTURITY

Having previously disassembled Beck's understanding of the material diffusion of risk and individualisation, it is now necessary to assess the cognitive effects of the twin dynamics on personal attitudes, values and perspectives. As a means of unpicking the argument, here we explore the assorted cognitive strategies used by individuals in making sense of risk. This discussion paves the way for a more sustained critique of the risk society account of reflexivity and a critical rethink of the relationship between experts and the public in contemporary society.

As noted in Chapter 6, the language of risk has become increasingly prevalent in the arenas of work, welfare and the economy. In recent decades, greater cultural emphasis has been placed on developing strategies of risk avoidance through personal planning and lifestyle choice (Lupton, 1999a; Giddens, 1994). Predictably, the swing toward personal responsibility for health feeds into heightened levels of risk consciousness. In this regard, the risk society perspective taps into the circulation of salient public discourses and values. Nonetheless, as has been noted, there remains insufficient analytical differentia- tion between risk as a remote possibility and as a likely occurrence. The fact that many western countries – including Britain, Germany, America and France – appear to be culturally fixated with risk does not signify that the probability of danger has risen for the average citizen. Whether risks to public health have increased in quantitative terms remains a topic of contest and debate. In the risk society thesis, the argument is pitched as one of quality rather than quantity (see Hinchcliffe, 2000: 136). Leaning on the three pillars of risk, Beck (1992; 1999) advises us that manufactured risks are comparatively more volatile, mobile and catastrophic than their predecessors, regardless of quantification and probability. Nonetheless, such skirting around the issue has not prevented Beck's contemporaries from questioning the extent to which risks to public health are escalating (Furedi, 1997; Giddens, 1998; Luhmann, 1993). For his part, Giddens is unwilling to equate heightened perceptions of risk with probability of harm:

> The idea of 'risk society' might suggest a world which has become more hazardous, but this is not necessarily so. Rather, it is a society increasingly preoccupied with the future (and also with safety) which generates the notion of risk. (Giddens, 1998: 27)

In a more committed vein, others have insisted that the current socio-political obsession with security exaggerates the possibilities of 'being at risk' (see Culpitt, 1999: 99; Furedi, 1997).

Whatever the arguments about the relative quantity of hazards in society, modern western societies certainly face increased risk awareness due to the expansion and diversification of social and scientific knowledge and the individualisation of risk. The general heightening of risk awareness within everyday culture is attributable to the broader development of scientific and technical knowledge. For example, medical advances such as pre-natal screening draw attention to complications in childbirth, raising sensitivity towards possible threats. As discussed in Chapter 6, the cultural trend toward planning and managing future possibilities is particularly pronounced in matters of personal health (Beck-Gernsheim, 2000: 124). Thus, public attitudes towards risk are critically linked to both wider cultural values and the objectives of social institutions. As Glasner (2000: 134) points out, 'any discussion of risk is as much about culture, institutions, perceptions, control and activity as it is about how risks are framed by experts'. As the capacity for amassing scientific knowledge has grown, so to has the cultural space open to public dialogue about risk. The more we hone the technical tools of identification, the more likely we are to unearth risk. This axiom underscores the imperfect connection between risk as possible and probable outcome:

> It is no accident that the risk perspective has developed parallel to the growth in scientific specialisation. Modern risk oriented society is a product not only of the perception of the consequences of technological achievement. Its seed is contained in the expansion of research possibilities and of knowledge itself. (Luhmann, 1993: 28)

Dominant understandings of risk are not created in a vacuum and must be framed within the ethos and aspirations of the age. With this point in mind, Furedi (1997: 6) argues that the current 'culture of fear' is wildly disproportionate to actual probabilities of danger. Beyond this, others have suggested that academic concentration on the dysfunctional aspects of the 'risk society' has contributed towards a wider moral panic (Culpitt, 1999). This claim may be overdone and Beck (1992: 75) is sharp enough to pre-empt the question frequently directed at the risk society thesis: 'is not the whole thing an intellectual fantasy, a canard from the desks of intellectual nervous nellies and risk promoters?' Naturally, Beck responds by reaching the opposite conclusion to Furedi. By downplaying, interpreting away and displacing risk, public health and future security are compromised. Whatever the lie of the land in this debate – and one suspects the middle ground to be the firmest – the ongoing advancement of scientific and medical technologies has unquestionably aided identification and forecasting of risk. Since we inhabit a culture preoccupied with managing upcoming events, it is expectable that we will foresee forthcoming threats and perceive ourselves to be more at risk. As we shall see, this future-oriented cultural trajectory creates sporadic conflicts about acceptable levels of harm and appropriate forms of risk management (Beck, 1992: 34; Caplan, 2000: 22; Lupton, 1999a: 3).

REFLECTING ON RISK:
TRADITIONS, TRADE-OFFS AND PLEASURES

Looking across the board, we can concur that the rising cultural presence of risk within the media, politics and the economy has engendered a relative growth in public concern. It is fair to assume that people are more cognisant of risk as a generic category than in previous ages. However, although such general statements may chime with the risk society perspective, they also camouflage significant contradictions and anomalies. While risk consciousness may be in the ascendant, the composition and quality of public understand-ings of risk remains disputable. Building on the findings of Chapter 5, in this section we flesh out the multiplicity of inputs which shape cognitive interpretations of risk, pinning down the problems that arise out of Beck's 'one size fits all' approach to reflexivity.

As discussed earlier, the risk society perspective differentiates between historical ages by sequencing forms of hazard attribution. During the evolution from pre-industrial to industrial societies, individuals move away from fate and religion as mechanisms of understanding and toward secular and rational perspectives. As cultures mature into the risk society phase, responsibility for risk is thoroughly humanised: 'risk society begins where tradition ends ... the less we can rely on traditional securities, the more risks we have to negotiate. The more risks, the more decisions and choices we have to make' (Beck, 1998b: 10). As we learnt in Chapter 6, there is more than a germ of truth in this argument. However, by adhering to a rigid historical framework, diffuse attitudes and values toward risk are unhelpfully compacted, both within and between eras. Beck's insistence on the contemporary individual as an exclusively rational assessor of risk forces him to neglect a host of culturally embedded features that shape vistas of risk. The rational actor of the risk society is uniformly vigilant and has no proclivity towards affective methods of interpretation. However, rather than adopting one-dimensional rational choice responses to expert knowledge, people will also be influenced by cultural, historical and emotional features. This means that public understandings of risk are more irregular and disordered than Beck allows for. Lay actors may 'follow logics that are obscure and apparently capricious, that can be encapsulated and "naturalised" in fatalistic beliefs, identities and senses of (non) agency' (Wynne, 1996: 53).

Regrettably, within the risk society narrative fatalistic understandings of risk are injudiciously sectioned to the pre-industrial era (see Beck, 1992; 1999: 50). In casting local customs and religions as the interpretative tools of 'traditional societies', Beck passes over the fact that people continue to be influenced by fate in patching together personal interpretations of danger (Cohn, 2000: 216; Vera-Sanso, 2000). Several studies into public perceptions of risk have indicated that fatalism acts as more than a simple reinforcement to rational choice (Eldridge, 1999; Reilly, 1999: 141). Take Douglas' study of attitudes towards Aids amongst sexually active homosexuals:

A large number of the community at risk are impervious to information; either they know unshakeably that they themselves are immune, or recognizing that death is normal they draw the conclusion that to live trying to avoid it is abhorrent. (Douglas, 1992: 111)

This caveat reminds us of the dangers of generalising the specific. In flagging potentially catastrophic risks as lightning rods of reflexivity, the risk society thesis misses the full attitudinal spread. For some, global risks may induce anxiety and critical assessment. For others, blame, attribution to fate and denial will be adopted as mix-and-match get-outs. Again, we can expect strategies of interpretation to mutate according to context. This said, it is probable that fatalistic modes of interpretation are likely to come to the fore in situations of extreme danger. This line of reasoning is illustrated by the response of members of the Oklahoma community to a tornado that struck the city in 1998.[2] In documentary interviews about the incident, a notable feature which emerged was the prevalence of fatalistic attitudes amongst those at risk. A number of interviewees chose to make sense of the disaster through religious reasoning or fatalistic attribution. Capturing the dominant mode of response to the tornado, one interviewee responded: 'You didn't have time to do anything, to think anything, you just had time to pray.' Of course, this merely

constitutes selective evidence, based around an extreme manifestation of risk. However, cases such as this do serve to remind us that,
in conceptualising catastrophic risk, communities may decide to
place their faith in religion or fate. Furthermore, it is quite possible
that the psychic absolution sanctioned by superstitions and rituals
can act as a safety valve for risk anxieties: 'There is, in other words,
a kind of defence mechanism for coping with the overwhelming
difficulty of living with inexplicable and uncontrollable, yet
emotionally important forces, which is to convert them into
identifiable agents, even superhuman ones' (Wynne, 1996: 54).

To be clear, to adopt a more inclusive position on public understandings of risk is not to suggest blind naivete on behalf of lay actors.
In negotiating risk, various coping strategies will be operationalised.
We need to retain the idea of lay actors as rational subjects, whilst
remaining receptive to the emotional, habitual and fatalistic influences
that steer individual interpretations of risk. Refuting the risk society
narrative, cultures do not possess internally cohesive frameworks of
sense making. The modus operandi of risk perception will vary
according to circumstance. Consequently, frames of risk reference
cannot be condensed down to an either-or choice between fate and
human volition. Instead individuals will draw across a range of
different cognitive strategies. In certain instances, lay actors may
hybridise interpretative prisms of risk. For example, fatalistic attitudes
towards risk may be 'half-believed' and supplemented by calculative
estimates (Giddens, 1999: 2).

In addition to religion and fate, risks to the self are managed through
ritual and performance. At the same time as the secularisation process
has eroded Christian faith in the West (see Turner, 1991) ceremonial
practices are still very much part and parcel of everyday life. Ritual
and performativity are particularly pronounced in relation to
maintaining a healthy body. Contemporary forms of body fetishism,
such as dieting, consuming herbal remedies, weight training, aerobics
and aromatherapy can all be conceived as modern rituals mobilised
to defend against ill health. To stretch the point, these customs may
fulfil a similar emotional function to religion, providing psychological insurance against risk.[3] Diet, health and fitness have become
something of an alternative to prayer and church-going. As Lupton
(1995: 4) muses, 'godliness' has been replaced by 'healthiness'. This
may overstate the case, but the popularity of ritualistic strategies of
risk avoidance does reopen the cracks in the risk society thesis. In
setting up critical and rational perspectives as conditioned responses

to risk, an unrealistic degree of perceptual unanimity is conveyed. Although lay actors are generally cognisant of catastrophic dangers, public attitudes will also be informed by varying degrees of pragmatism born out of lived experience. Redeploying Marshall's work on social class, such a dualistic attitude towards risk might be described as 'informed fatalism' (Marshall, 1988: 143). Evidently, strategies of avoidance are not mobilised at the first sign of possible danger (Reilly, 1999: 131). Under certain social circumstances lay actors can be remarkably tolerant to the threat of risk. Indeed, an interesting – and largely untapped – area of research revolves around public attitudes towards manufactured risks, when set against potential gains (see Douglas, 1985: 59; Vera-Sanso, 2000: 126). In some cases, manufactured risks may be accepted as the quid pro quo for tangible benefits and social progress (see Irwin et al., 2000: 95). In as much as chemical, nuclear and genetic technologies are productive of certain risks, they may also be seen to advance future quality of life. Indeed, techno-scientific cautiousness begets its own hazards. Without risk taking in science and medicine, a number of life-saving technological inventions would not have materialised (Goldblatt, 1995: 175). Unfortunately, Beck has little to say about risk trade-offs or the relationship between risk taking and social development. Although these modalities do not fit comfortably into the risk society framework, they remain vital areas of risk psychology and reasoning.

In order to construct a holistic understanding of the relationship between risk, public perceptions and lived experience, the full gamut of cultural attitudes needs to be considered. Despite Beck's dystopic account of risk as a universally negative phenomenon, it is clear that risk taking can act as a funnel for excitement and adventure: 'think of the pleasures some people get from the risks of gambling, driving fast, sexual adventurism, or the plunge of a fairground roller-coaster' (Giddens, 1999: 2). In recent years, the interconnectivity between risk and pleasure has been highlighted by the increasing popularity of extreme leisure pursuits, such as skydiving, snowboarding and white-water rafting (Lupton and Tulloch, 2002a: 114). Dangerous activities which have a sensuous texture may make the nectar of risk sweet for some. Lyng (1990) reads cultural preferences for voluntary exposure to risk as a desire to engage in 'edgework'. Of course, historically speaking, various forms of edgework have been carved into youth subcultures. From Teddy Boys and Mods, to Ravers and Boy Racers, risk-taking behaviour has acted as a springboard to social status and a means of reinforcing identity. Although 'edgework' pertains to an encased set of risky practices (Lupton, 1999a: 113), Beck's overemphasis on risk aversity neglects subcultural performances that utilise risk taking as a technique of fulfilment and a mechanism of integration: 'Voluntary risk taking is often pursued for the sake of facing and conquering fear, displaying courage, seeking excitement and thrills and achieving self-actualisation and a sense of personal agency' (Lupton and Tulloch, 2002a: 115).

Despite evidence of autonomous risk-taking activities, in the risk society thesis the negativity of risk is accentuated by constant reference to the dangers and pathological features of modern life (Goldblatt, 1995: 175). Thus, the complex association between risk, pleasure and desire is strikingly absent (Culpitt, 1999: 113; Irwin et al., 2000: 98). Hence, arguably the most striking imperfection in the risk society thesis is a refusal to recognise the diversity of hermeneutic approaches which people employ in their routine encounters with risk. In opposition to the universalism inherent in the risk society argument, a dense network of habits and dispositions will influence the way people respond to risk in the course of everyday life.

RETHINKING REFLEXIVITY: FACTORING IN THE AESTHETIC

Having alluded to multiple modes of risk interpretation, I now wish to spell out the consequences of a utilitarian approach to the subject for Beck's portrayal of reflexivity. We begin by collecting up the neglected aesthetic aspects of reflexivity as a means of pointing up the absent layers in the risk society narrative. Siding against Beck, we go on to advocate a subtler understanding of flows of risk knowledge and a more sophisticated conception of the lay–expert relationship. We have previously noted that the risk society perspective is short on appreciation of the cultural milieu in which day-to-day sense making occurs. Opposing an excessively rational notion of the self, it has been argued that everyday attitudes towards risk are structured by a range of emotions, values and beliefs. In the remaining sections of the chapter, I wish to propose an understanding of reflexivity which builds upon these guiding principles. In the course of discussion, I lean upon the work of Scott Lash and Brian Wynne, two theorists who have been instrumental in tracing the cultural and aesthetic feeders of reflexivity.

Lash (1993; 1994) and Wynne (1992; 1996) are agreed that Beck overplays public dependence on scientific and technical information in the formation of reflexivity. Despite following different routes of critique, both theorists concur that Beck's presentation of reflexivity is anchored in the power of macro institutional structures rather than micro forms of public agency. In recounting the insurgence of public reflexivity, Beck prioritises the information-bearing role of institutional experts and counter-experts within science, government, media and the legal system. From Lash's corner, this overriding concern with social structure and institutional rationality causes Beck to neglect the aesthetic and cultural drivers of reflexivity (Lash, 1994: 201). In many respects, the risk society argument follows the trajectory of Weber's theory of social action, with epochal changes being related to transitions in human behaviour. For Lash (1993: 2), the value-rigidity of the risk society model forces Beck to reproduce a 'one-sided notion of contemporary subjectivity'. Arguing against such reductionism, Lash maintains that reflexivity is multilayered and must be understood in relation to a sweep of cultural practices and behaviours. Unfortunately, the risk society thesis fails to address the power of both tradition and aesthetics in moulding responses to risk.

Secondly, Lash disagrees that techno-scientific modernisation inexorably undermines the credibility and functions of social institutions. In spite of suffering episodic challenges and confrontations, expert systems appear to be transmuting rather than dissolving (Lash and Urry, 1994: 119). According to Lash, social institutions are undergoing a sustained period of restructuring, with modern communication systems succeeding traditional influences. Although this position is not too far removed from the ground occupied by Beck, Lash moves an important step further, delving into the wider expansion of symbolic representation. As discussed in Chapter 4, the media is a very particular type of 'structure'. As far as the social construction of risk is concerned, technological diversification, consumer demands and the trend toward narrowcasting have led to a discernible shift from information to entertainment-based programming. With the introduction of new technologies such as the internet, digital television and virtual reality, styles of public interaction are shifting from mass engagement to more individualised forms. For Lash, the broader aestheticisation of culture through consumerism, travel and tourism indicates a rise in the role of the visual and a recession of the informational: 'For example, the increasingly reflexive nature of economic growth is aesthetic, as products are increasingly associated with images; as symbolic intensity at work often takes the form of design rather than cognition' (Lash, 1993: 19).

With specific reference to risk, the import of the mediated aesthetic is well supported by audience reception studies. For example, Corner et al.'s seminal (1990a; 1990b) analyses of the mediation of environmental risk uphold the value of cultural and symbolic features in structuring meaning. Further, Harris and O'Shaughnessy (1997) observe that public understandings of the BSE crisis were influenced by a nucleus of core images – lurching 'mad' cows, incinerated cattle and bloodstained slaughterhouses – which came to represent the crisis in the eyes of the public. Subsequent film-footage of the disorientated motions of human vCJD sufferers only served to rekindle public anxieties. More recently, haunting footage of the terrorist attack on the Twin Towers needled into visceral and emotive veins, rather than rational forms of reasoning. Backing Lash, the operation of the symbolic in fuelling personal interpretations of risk urges a more refined understanding of reflexivity. Given the expansion of visual representations of risk, Beck's normative version needs to be supplemented by an appreciation of aesthetic and emotional dimensions of reflexivity.[4]

> We need to move from a 'thin' idea of citizenship, where a disembodied, rational self judges proposals and arguments on the basis of abstract rationality, to a 'thicker' account which emphasises the cultural work that stands behind the achievement of such a 'citizen identity'. (Szerszynski and Toogood, 2000: 222)

REFLEXIVITY, TRUST AND CULTURAL KNOWLEDGE

In fairness to Beck, it would be unjust to suggest that the risk society thesis is oblivious to the aesthetic dimensions of reflexivity. On occasion, Beck does tinker around the aesthetic fringes (1992: 24; 1995a: 100). Nonetheless, vacillation between rational and cultural poles leaves the risk society thesis in something of a pickle. It would seem that Beck wants to maintain that reflexivity is a rationally clean and politically stimulating activity, whilst sneaking cultural and aesthetic factors through the back door. Lining up with Lash, Wynne (1996) criticises the precedence accorded to institutional rather than experiential facets of reflexivity. For Wynne (1996: 47), Beck's rationalistic construction of lay reflexivity implies an institutional and contractual model of social action. Instead of responses to risk being exclusively generated out of rational engagement with the relations of definition, social actors will construct semi-autonomous, culturally independent understandings. The intrinsic problem with the risk society theory is that, at root, public reflexivity amounts to little more than choosing between (counter-)expert versions of truth. In actuality, public understandings of risk are formulated with, without, and sometimes *in spite of* the information provided by expert bodies (Lupton, 1999a: 110). While the risk society perspective accentuates the role of institutions in generating risk meanings, it is clear that lay actors run into expert claims within a wider social context. Reflexivity is not executed within the confines of cognitive-rational relationships between individuals and institutions. Instead, risky events are culturally interpreted in conjunction with a range of embedded beliefs, values and practices. The 'sense' that is made of risk is both situated and cultured.

Wynne's criticisms of the risk society thesis are anthropomorphised in an outstanding ethnographic study of Cumbrian sheep farmers (1992; 1996). After the Chernobyl incident, concerns were raised that leaked radiocaesium could be carried in the air from the Belarussian nuclear plant and deposited on British soils. Government scientists were quick to scotch these claims, arguing that the radioactive caesium isotopes deposited as fallout from Chernobyl would be washed off vegetation into the soil where it would be chemically 'locked up'. Nevertheless, farmers in Cumbria became increasingly concerned about the movement of pollution from vegetation to sheep and on through the food chain. Scientific experts conceded that this movement was hypothetically possible, but insisted that – even if sheep became contaminated – the infection would only endure for three weeks. At the time, government scientists stated that there would be no future contamination of livestock after the initial flush, with the risk of contamination receding after a three-week quarantine period (Wynne, 1996: 53). Thus, a temporary ban was placed on the movement and slaughter of sheep in the region (Wynne, 1992: 283). Toward the end of the quarantine period, the government stunned the local community by announcing that the ban on livestock trade would be extended indefinitely. It soon became apparent that the scientists had located abnormally high levels of contamination in the Cumbrian soils, which they duly attributed to the Chernobyl reactor explosion. The validity of this claim was widely disputed by local farmers, who believed that much of the contamination in the soils had not come from Chernobyl but from a more proximate source, namely the Sellafield nuclear plant. Despite denying this possibility at the time, months later in a 'leaked memo' it was revealed that scientists had – rather diplomatically – traced half of the radiocaesium in the soil to Chernobyl and the remaining half to 'other sources', including fallout from weapons testing and emissions from the Sellafield plant (Wynne, 1996: 65).

In some respects, Wynne's case study speaks in favour of the risk society thesis. The hill farmers observed in Cumbria were not bound into relationships of absolute trust with expert systems. In line with Beck's culture of reflexivity, the native population were consistently sceptical about governmental findings: 'outbursts of frustration at the experts' ignorance occurred often' (Wynne, 1996: 66). Local people were quick to identify inconsistencies in expert arguments and challenged what they perceived to be deliberate strategies of deflection. Sharing the seat with Beck, Wynne argues that public scepticism and a broader suspicion of expert knowledge are becoming increasingly common features of modern life. However, the two theorists part company when it comes to the sourcing of information which fuels public reflexivity. The risk society thesis suggests that investigation by (counter-)experts is the medium through which risks to health are identified. Using the mass media as a mouthpiece, various experts publicise and contest the evidence. The conflict between institutional actors and oppositional groups is played out in front of an informed and interested public, who seek to draw out a reliable version of events. In contrast, Wynne's research questions the assumption that public understandings of risk are constructed on the basis of information provided by external experts: 'public experience of risks, risk communications or any other scientific information is never, and never can be, a purely intellectual process, about reception of knowledge *per se*' (Wynne, 1992: 281). Conversely, the Cumbrian example illustrates that informed understandings of risk can be located in the lay-public domain outside of the parameters of expert or counter-expert circles. Lay actors are capable of accumulating substantial intellectual knowledge about risk through personal observation and lived cultural practices. Far from being encouraged by the counter-expert discourses of oppositional groups writ large in the media, Wynne's (1992: 292) participants drew upon tacit knowledge and cultural experience. The Cumbrian farmers mistrusted external scientific knowledge, favouring local networked knowledge and collective sensibilities. In Wynne's study, local farmers were keen to uphold their own historical experience of cultivating the land as a firmer basis for informed decision making than the one-off studies of scientists. As a coda, this denotes an interesting twist in terms of our understanding of the lay–expert nexus. In the Cumbrian case, the boundaries between 'the experts' and the 'lay actors' were decidedly blurred. Indeed, in the final analysis, the most accurate body of risk knowledge was provided by the local farmers, not the government or their scientific experts. Recast in this light, ignoring expert advice can make perfectly sound and 'rational' sense. Unfortunately, the risk society thesis tends to overlook culturally formulated knowledge and exaggerates the value of institutional/scientific as against local sources of information. Moving a step beyond, in contrasting scientific 'facts' with cultural 'values' we run the risk of attributing false impartiality: 'It is not a matter of lay public "cultural" responses to "meaning-neutral" objective scientific knowledge, but of cultural responses to a *cultural* form of intervention – that is, one embodying particular normative models of human nature, purposes and relationships' (Wynne, 1996: 68).

In addition to questioning binary representations of science and culture, Wynne's study also raises absorbing issues about the quality of public trust in expert systems. The sheep farmers under observation were involved in ambiguous trust relations with institutional experts. Insofar as respondents questioned informational inconsistencies, they were also aware of their ultimate dependency on expert decisions. For many of the farmers, this led to behaving 'as if' the experts were trusted (Wynne, 1996: 65). Thus, most farmers conceded to institutional authority even when they were sceptical about the advice provided. As one respondent mordantly commented: 'You can't argue with them because you don't know – if a doctor jabs you up the backside to cure your headache, you wouldn't argue with him, would you?' Such ambivalent behaviour towards expertise can be described as 'virtual trust' (CSEC Report, 1997).

The layered and contingent constitution of trust shows up the rather reduced notion of trust employed in the risk society narrative. For Beck, public trust is depicted as an absolute commodity, which institutions either have or lack. On the contrary, recent studies have demonstrated that public trust in institutions is much more than an affirmative-negative equation (CSEC Report, 1997; ESRC Report, 1999). Rather than being amenable to the non-aligned theorising of the risk society thesis, trust relations are inextricably linked to particular cultural and historical conditions. As Jasanoff (1999: 150) notes, 'judgements about the nature and severity of environmental risk inevitably incorporate tacit understandings concerning causality, agency, and uncertainty, and these are by no means universally shared'. Countries with embedded traditions of institutional openness are less likely to suffer spectacular declines in public trust (Cohn, 2000). As indicated in Chapter 3, the actual constitution of trust will vary between cultures. Trust as a category is made up of a series of composite factors, such as authority, perceived ability to act, previous competence and informational credibility. As with risk, we can expect particular dimensions of trust to be accentuated and reduced in different spaces and places (Allen, 2000: 24).

Returning to the risk society model, it is clear that the partition erected between trust relations in industrial modernity and the risk society is rather rickety. Forms of trust and levels of reflexivity cannot be meaningfully conceptualised in an unconditional fashion. In the same way that agents in contemporary western cultures are not universally critical of expert systems, industrial societies cannot be seen as havens of absolute public trust. Opposing the ingrained stiffness of the risk society perspective, public distrust predates the onset of the risk society. Throughout modern history, the power of expert institutions has been a bone of public contention:

> The assumption of lay public trust in expert systems under conditions of so called simple modernity has to be replaced by a more complex notion of this relationship, in which ambivalence is central and trust is at least heavily qualified by the experience of dependency, possible alienation, and lack of agency. (Wynne, 1996: 52)

Public disenchantment with the functioning of expert systems has deeper historical roots than Beck is prepared to acknowledge (Culpitt, 1999: 119). A lack of visible opposition to public institutions – in the past or the present – should not be translated as lay satisfaction with the performance of expert systems. To act implies being *able* to do so. As discussed in Chapter 6, being able to act is not determined solely by individual choice, but is instead dependent upon facilities, tools and technologies (Lodziak, 2002: 97). Therefore, a lack of direct opposition towards expert systems cannot be comfortably equated with unqualified public trust (Wynne, 1996: 50). In the risk society – as in industrial modernity – access to material resources is a prerequisite of reflexive engagement: 'Although in principle expert knowledge is accessible to everyone, if we have the time and money to acquire it, in practice most of us can become experts in only one or two areas of the expert world' (Mol and Spaargaren, 1993: 451).

Where do these objections leave us in terms of our understanding of relationships between experts and the public in contemporary society? Firstly, it is evident that the terms 'expert' and 'public' grant descriptive utility but lack analytical purchase. The risk society thesis depicts experts and the public as two relatively homogeneous and polarised groups. This coarse all-purpose approach generates an exaggerated version of social relations in contemporary western society. Working with such ideal type categories hides the diversity of opinion within and between groups and, in so doing, conceals as much as it reveals (see Kerr and Cunningham-Burley, 200: 293; Mythen, 2002). As Macgill (1989: 50) notes, 'simply identifying "the public", is not to fix on a uniform mass of fossilised opinion ... there is not a simple homogeneous pattern of opinion. On the contrary, there are striking differences'. It needs to be acknowledged that the term 'lay public' houses a myriad of overlapping and distinct sub-populations and incorporates a diverse range of attitudes towards risk. Such demographic complexity is not adequately acknowledged in the risk society thesis, which represents – and, arguably, is representative *of* – a distinctly selective public. In as much as we can identify a comparatively reflexive, ecologically concerned and politicised sector, one public should not be taken as expressive of another: 'what we are just beginning to realise is that there are many "publics" in society, and that any given individual may move in and out of a number of bonding groups' (CSEC Report, 1997: 19).

In a similar vein, expert institutions can be expected to accommodate greater diversity of ideas and opinions than the lay–expert binary suggests (Cottle, 1998; Irwin, 1989: 30). The lay–expert relationship encapsulates too wide a range of subject positions and identities than can be realised in a single split, with the boundaries between expert and lay groups being indistinct and permeable.[5] To argue otherwise is to write off important cultural patterns for the sake of maintaining an orderly theory. Of course, this is not to pretend that general tensions do not exist between the public and risk-regulating institutions, nor to suggest an egalitarian relationship between the two parties. Clearly, the lay–expert relationship is still shot through with power differentials. However, we do at least need to recognise the room for manoeuvre within and between each position. This suggests that a degree of fluidity needs to be built into our understanding of trust relations in contemporary society. Rigid lay–expert groupings need to be recast as 'liminal categories' with interfacing boundaries (Crook, 1999: 174).

CONCLUSION

Before coalescing our key findings, it is worth reflecting on the broader developmental curve which has taken shape over the last three chapters. In Chapter 5, we first cast an eye over public perceptions of risk, comparing and contrasting the findings of empirical studies with the theoretical arc of the risk society thesis. Although our discussion remained formative, it became clear that Beck's argument skates over important cultural influences – such as gender, class, ethnicity and place – which anchor individual interpretations of risk. In Chapter 6, this observation was substantiated in relation to routine material encounters with risk within the family and the workplace. Training our lens on individual–institutional interactions, we also examined the cultural diffusion of individualisation. In this chapter, we have moved from the material to the abstract/symbolic, prioritising the impacts of risk on the formation of personal reflexivity and the constitution of institutional trust. Spanning these three chapters our analysis has turned up a number of crucial political issues which have purposely been left hanging. These questions which congregate around the 'politics of risk' will be attended to and resolved in the final chapter.

Over the course of the book, it has become apparent that responses to risk are housed along a continuum, ranging from fatalistic acceptance to political opposition. In this chapter, we have demonstrated that growing cultural recognition of risk does not mechanically produce public aversity. Some distance away from global catastrophe, the pleasures of certain risks can act as a vent for sexual desire, self-identity and personal fulfilment. For young people inured to the insecurities and demands of biography building, blanket intolerance of risk is atypical. Instead, it would seem that a more conditional attitude prevails. On the one hand, 'healthy bodies' are protected and nourished through exercise, diet, vitamin supplements and practices of beautification. On the other, 'risky bodies' are exposed to danger through interfaces with risky products and practices. The continued prevalence of alcohol consumption, drug taking, unsafe sex and extreme sports within youth cultures is indicative of the pleasures derived from risk. The risk society thesis wrongly assumes that cultural advertisement of risk promotes unilateral disquiet and strategies of avoidance (see Lupton and Tulloch, 2002b). It should be remembered that, in an uncertain world, moderation and protection do not ensure well being:

People often develop a fatalistic attitude towards risk because they have observed that life does not always 'play by the rules'. Someone who drinks heavily and smokes may live to a ripe old age, while an ascetic non-smoking jogging vegetarian may die young. (Lupton, 1999a: 111)

In this sense, reflexivity is a distinctly social construct which grows out of cultural interactions as well as expert information. As we have seen, Beck's heavy inflection on the institutional/structural dimensions of reflexivity has stirred a series of objections (Alexander and Smith, 1996; Lash, 1994: 201; 2000: 50). It would appear that the risk society approach to reflexivity is destabilised by its bimodality. Beck wants to maintain that public reflexivity grows against expert systems, whilst simultaneously insisting that lay actors remain information-ally dependent upon scientific and intellectual knowledge (Cottle, 1998: 13). In modelling the individual as rational and goal-oriented, Beck overlooks the situated fashion in which people live, work, love and play. In the risk society thesis, individuals relate to relationships and surroundings in an instrumental and utilitarian manner, rather than engaging in 'creative social action which is structured in terms of cultural forms' (Strydom, 1999: 48). This utilitarianism serves to decontextualise the habitus in which agents make sense of risk in their everyday lives. As has been illustrated, abstract systems of risk definition tend to be mediated through local structures (Lash and Wynne, 1992; Wynne, 1992). In formulating knowledge about risk, lay actors are more likely to trust the proximate opinions of friends, colleagues and family than those of 'outside' experts (Caplan, 2000a: 22; Marris and Langford, 1996: 37). In conclusion, responses to risk cannot be anything other than culturally grounded phenomena. This suggests that a crucial interregnum may exist between information about risk and precautionary action. What intervenes, it seems, is culture itself – understood as a set of everyday practices through which meaning is generated. As Beck envisages, in an ideal world, collective energies could be focussed on eliminating poverty and environmental risks. However, in reality, the accessibility of resources – such as time, money, energy and technology – will determine paths of action available to the individual. Even in the affluent West, the realm of necessity and associated tasks of self-reproduction eat up much of the time and energy available for emancipatory activities (see Lodziak, 1995; 2002).

It has also been demonstrated that the rudimentary understanding of trust relations within the risk society thesis amplifies the divide between experts and the lay public. As we shall see in Chapter 8, the degree of political dissent between institutional actors and non-expert individuals is not easily calibrated. Although attributing blame to experts has become an increasingly common response to risk incidents, it is probable that Beck overplays levels of institutional distrust. In depicting public critique as a habituated response to social dangers, the situated context in which individuals make decisions about risk is lost and jagged and variable trust relations are simply carpeted over.

8
The Politics of Risk

Having considered the ways in which risk impacts upon social structures and everyday experience, we are now in a suitable position to extend our investigation into the relationship between risk, reflexivity and political engagement. Of course, the politics of risk has been a constant undercurrent. In Chapters 2 and 3, the politics of environmental risk production were approached through analysis of the functions of science, technology and the state. In Chapter 6, we touched upon the political waves created by structural transformations in work, the family and welfare. Latterly, in Chapter 7 the linkages between risk, trust and political reflexivity were broached through the cultural critiques of Scott Lash and Brian Wynne. Thus, the affiliation between risk and politics is an issue which has bubbled under the surface throughout. In this, the final chapter, we directly confront the political dimensions of the risk society thesis. More explicitly, we will be interrogating Beck's claim that a 'reinvention of politics' is underway, altering the nature and scope of public engagement. To this purpose, evidence of a discernible move towards subpolitics will be evaluated with an eye to current trends and germane criticisms of the risk society approach (Culpitt, 1999; Hinchcliffe, 2000; Lupton, 1999a).

As is the fashion, it is first necessary to sketch out the political aspects of the risk society perspective, drawing upon the arguments outlined in *Risk Society* (1992), *Ecological Politics in an Age of Risk* (1995a), *The Reinvention of Politics* (1997) and *Democracy Without Enemies* (1998a). This synopsis will provide the raw material for consequent assessment of Beck's understanding of the relationship between risk, reflexivity and politics. In later sections, evidence for and against a marked shift in the content of political debate will be documented. To tie down abstract theory, Beck's notion of reflexivity will be related to the current political controversy surrounding genetically modified foods. A primarily positive application of the risk society thesis will subsequently be counterbalanced with reference to the Foucauldian critique. Drawing upon discourse theory, the restrictive potential of expert knowledge and language will be considered as a means of teasing out the politically repressive uses of discursive formations of risk. In conclusion, we rejoin the concept of subpolitics as a means of engaging more rigorously with the effects of manufactured risks on political organisation.

THE DEATH AND BIRTH OF POLITICS

In *The Reinvention of Politics* (1997), Beck avers that a rudimentary shift in the locus of political decision making has occurred in the last half a century. In effect, political decision making has migrated from systems of national governance into economic, technological and scientific domains. As a result, major social decisions are no longer made by elected political representatives. Instead, an elite band of unaccountable and unelected scientists, business leaders and legal specialists are responsible for fixing the boundaries of acceptable technological development. For Beck, a lack of visible responsibility for manufactured risks obfuscates important political issues and leads to public frustration. Within the present formal democratic system, public involvement is restricted to a superficial choice of political representatives, alongside hierarchically organised 'consultation' about the constitution of political programmes. It is important to stress from the outset that Beck does not seek to provide a systematic critique of the formal democratic process.[1] Instead, the connections between risk, public concerns and political change are given precedence.

The risk society thesis posits that the pace of techno-scientific change has resulted in governments assuming a reactive rather than proactive position on social risks. It has been scientists, technologists and multinational companies who have driven 'advances' in genetic, nuclear and biochemical technology, whilst national governments have assumed a back seat, assenting to market forces (Beck, 1995b; Ho, 1997). In his later work, Beck (1999; 2000a) fingers the globalisation of capital as the dominant force behind the transference of political power:

> During the first age of modernity, capital, labour and state played at making sand cakes in the sandpit (a sandpit limited and organised in terms of the nation-state) and during this game each side tried to knock the other's sand cake off the spade in accordance with the rules of institutionalized conflict. Now suddenly business has been given a present of a mechanical digger and is emptying the

whole sandpit. The trade unions and the politicians on the other hand, who have been left out of the new game, have gone into a huff and are crying for mummy. (Beck, 2000a: 89)

In industrial society, lay publics came to accept that crucial social decisions would be made by elected governments through legislation, on the basis of expert knowledge. In the risk society these decisions have been hijacked by science and business and no longer fall within the jurisdiction of the state. Obviously, this argument is far from exceptional. It is commonly recognised that globalisation has left significant marks on the sovereignty of national governments and redrawn traditional political boundaries (see Held, 1995; McGrew, 2000; Waters, 1995). However, as always, Beck wishes to push the argument under the capacious umbrella of risk. What is important, as far as the risk society thesis is concerned, is that the very institutions which the public turn to for guidance on major political issues no longer have control over the decisions which could ensure safety. As a result, issues of accountability and responsibility creep to the fore:

Neurotechnologies and genetic engineering are reshaping the laws that govern the human mind and life. Who is doing this? Technological experts? Medical experts? Politicians? Industry? The Public? Ask any of them, and the reply will be the one Ulysses gave the cyclops: 'nobody'. (Beck, 1995b: 505)

Thus, the political problematique arises out of a lack of active democracy in large-scale decision making about economic and technological development. In the first instance, a minute number of specialists are involved in taking scientific and technological decisions. Secondly, major decisions about risky technologies often bypass the parliamentary process, being enacted 'in the twilight zone' where science and industry merge (Beck, 1995b: 506). Sine qua non, national governments find themselves having to legitimise decisions they did not effectively 'take' in the first place. At present, a plethora of inherently social issues slip the net of the formal democratic process. By and large, developments in genetic technology, microelectronics and nanotechnology are resolved by industry and science, with governments undertaking little more than a rubber-stamping exercise. For Beck, an apparent lack of political responsibility leads to a fait accompli in which risks are 'produced by industry, externalized by economics, individualized by the legal system, legitimized by the sciences and made to appear harmless by politics' (Beck, 1998b: 16).

The loss of state power can be traced back to the changing infra-structure which globalisation facilitates, with political conflicts projecting risk from the local to the global. In harmony with Roland Robertson (1992), Beck (1999: 15) notes the increased prevalence of 'glocal' issues which cannot be solved by top-down national politics.[2] In the risk society narrative, the globalisation process sends politics in two different directions. On the one hand, there is evidence of an ongoing form of 'globalisation from above' through international treaties and the dictums of global political elites. Conversely, the diversification of politics also stimulates 'globalization from below' through the collective actions of groups acting outside of the formal democratic arena (Beck, 2000b: 37).

SUBPOLITICS IN THE RISK SOCIETY

In *Risk Society* (1992) Beck blows the trumpet for 'globalization from below', championing a system of differential or 'subpolitics' in which politics becomes generalised and centreless (Beck, 1992: 227). Through self-coordination and direct action citizens can contest vital issues affecting the environment, science, business and education (Beck, 1998a: 152). Since the publication of *Risk Society* (1992), Beck has continued to advocate subpolitics as a progressive form of public involvement which enhances democracy in western cultures. In *World Risk Society* (1999), the increasing popularity and power of global social movements such as Greenpeace, Amnesty International and Terre des Hommes are vaunted. Beck believes that the successful direct actions taken by NGOs highlight the failure of national parliamen-tary systems to respond to pressing political affairs and to generate positive social change. The rapid rise of subpolitical movements raises the possibility of a more deliberative and inclusive form of democracy through which environmental risks could be regulated:

> Subpolitics means 'direct' politics – that is, ad hoc individual par-ticipation in political decisions, by-passing the institutions of representative opinion formation (political parties, parliaments) and often even lacking the protection of law. In other words, subpolitics means the shaping of society from below. Economy, science, career, everyday existence, private life; all become caught up in the storms of political debate. But these do not fit into the

traditional spectrum of party-political differences. What is characteristic of the subpolitics of world society are precisely *ad hoc* 'coalitions of opposites' (of parties, nations, regions, religions, governments, rebels, classes). Crucially, however, subpolitics sets politics free by changing the rules and boundaries of the political so that it becomes more open and susceptible to new linkages – as well as capable of being negotiated and reshaped. (Beck, 1999: 40)

In the subpolitical model, traditional political affiliations of party, class, gender and ethnicity become outmoded – politics is no longer about acquiescence in grand narratives of liberalism, conservatism or socialism. In contrast, subpolitics offers a more direct route to political enagagement: 'one can spare oneself the detour through membership meetings and enjoy the blessings of political action by heading straight to the disco' (Beck, 1998: 170).

Thinking outside of the risk society box, designs for political restructuring have been common within social theory, particularly within the Marxist tradition. Despite sharing some common ground with Marx – for example, over the inexorability of capitalist crisis and the inevitability of political opposition – Beck disputes the claim that an identifiable class will spearhead political revolution. Whereas pure Marxists maintain that political revolt will be conducted by the exploited working class, Beck believes that subpolitics is a classless and inclusive form of political action: 'Of course, everybody asks who is the political subject of risk society ... my argument is as follows: nobody is the subject and everybody is the subject at the same time' (Beck, 1998: 19).

The universality of social bads means that actors from diverse backgrounds come together in the subpolitical space, 'to reinvent the co-ordinate system and to reset and realign the switches' (Beck, 1998a: 104). A similarly decentred version of political change can be found in the work of André Gorz (1982; 1998; 1994; 2000). For Gorz, rather than the traditional working class, the subjects of political revolution will be an indistinct 'non-class of non-workers'. Whilst Beck and Gorz are agreed on the need for political reformation, the two diverge in their interpretations of the motors of change. Where Gorz stresses the crisis of the employment system, Beck zooms in on the proliferatation of manufactured risks. In the German context, Beck's project has attracted comparison with Habermas's work on the public sphere (see Prior et al., 2000). Meanwhile, in Britain, the Giddensian concept of 'life politics' bears more than a passing resemblance to subpolitics.[3] For Giddens (1994: 14), life politics is about small-scale local activities which take place outside of the formal political system, free from delineated hierarchies. As with subpolitics, engagement in life politics is characterised by debate about future-oriented ethical issues. Sharing overlaps with such theorists, Beck strives to envisage a more deliberative and inclusive polity. In the risk society thesis, the scale and extent of manufactured threats force citizens into socio-political reflection. However, risk concerns cannot be satisfactorily expressed through the traditional routes of the formal political system. Thus, the diffusion of risk undermines traditional power bases, making society susceptible to political restructuring. Consequently, scattered pockets of subpolitics materialise, offering an active and meaningful alternative to the formal party process. For Beck, the deleterious effects of short-term economic goals and blind technological development give rise to the possibility of constructing a new global political order along the lines of a 'cosmopolitan democracy' (Abbinnett, 2000: 115).

THE GM FOOD DEBATE: SUBPOLITICS IN PRACTICE?

Having provided a rough outline of Beck's hypothesis, it is now time to evaluate the emancipatory possibilities of subpolitical engagement. In this section, the controversy that has arisen in response to the development of genetically modified organisms will be mobilised as a means of considering both the potential power and the current coverage of subpolitics. A preliminary discussion of risk as an enabling political force will be counterpoised by the restraining possibilities attached to risk in the discursive approach. In synthesis, we reflect on the relative value of each perspective in making sense of contemporary social and political trends.

In concordance with Beck, a cursory glance over the terrain of European politics indicates relatively low levels of public involvement in the formal democratic process. If voting statistics are reliable indicators of public interest, many people are sceptical about the ability of the political system to effect positive social change (Brynner and Ashford, 1994; Park, 1996).[4] A mood of general dissatisfaction with formal politics appears to be particularly prevalent amongst young people. However, recent political demonstrations – against globalisation, the Gulf conflict and transnational corporations – demonstrate that many young people remain committed to political causes. Further, the increasing visibility of older demonstrators on various campaign marches indicates that the 'space' of engagement may be drifting away from the formal system and toward direct action and protest.[5] Underlying shifts in political activity are also highlighted by the rising profile of Non Governmental Organisations (NGOs) which work both inside and outside the boundaries of the formal political arena. In the twentieth century, the number of NGOs rose from around 200 in 1909, to over 50,000 by the year 2000 (Held et al., 1999: 151). The single issue campaigns waged by various NGOs have impacted upon the structure of politics, contributing towards the development of novel ways of enhancing public involvement, such as citizens' juries, deliberative polls and consensus conferencing (see Barnes, 1999; Coote and Lenaghen, 1997; Coote and Mattinson, 1997; Smith and Wales, 2000). As a reaction to public discontent about the unresponsiveness of social institutions, a number of quasi-autonomous government bodies have been set up to plug the gap between citizens and the state.[6] Although these developments do indicate shifts in political practices, the diffusion of subpolitical activities is extremely difficult to quantify. To gain insight into transitions in the nature of political engagement, we are perhaps better served employing a more qualitative approach. Taking account of the political controversy which continues to surround GM foods, it may be fruitful to employ this issue as a touchstone for debate.

In theory, genetically modified foods bear all the hallmarks of manufactured risk (see, Beck, 1999: 107; Wales and Mythen, 2002). In the first instance, the potential risk is created by human endeavours within technology and industry. Second, given that base altered foodstuffs are used in many products, the diffusion of genetically modified foods is difficult to regulate. Third, the possible effects of genetically modified foods on the human body and the environment are unknown. Fourth, the risk presented by GM foods is effectively boundless and potentially catastrophic.[7]

Genetically modified food crops were first commercially grown in the United States in 1995 and have since been developed in many other countries, including Britain, Mexico and Brazil. In 1996, genetically modified soya and maize were first sold by Monsanto, an American multinational company. It has since emerged that the company routinely mixed GM and non-GM crops, making it impossible to differentiate between genetically modified and GM-free products (ESRC Report, 1999: 9). GM crops are already being grown on over 35 million hectares of land around the globe (Giddens,

1999: 5). Currently, a wide variety of genetically manufactured foodstuffs appear on the market, ranging from oilseed rape to fruit and vegetables. Those in favour of genotechnology argue that GM foods are more flavoursome, resistant to damage and amenable to storage for extensive periods: 'Like previous technological innovations, it holds out the promise of cornucopia: the end of food shortages and world hunger, poverty and disease, weather and season dependence' (Adam, 1998: 11).

Despite the claims of its proponents, the release of GM foods into the human food chain has provoked widespread public concern in many countries around the globe. A number of NGOs such as Greenpeace and Genewatch have challenged food manufacturers and governments, arguing that GM foods may present a threat to public health and could cause long-term damage to the environment. Academic critics, including Beck (1999: 105), have argued for implementation of the precautionary principle, pointing out that the long-term physiological effects of consuming genetically modified foods are incalculable.

Technically speaking, the genetic modification of food involves the isolation of a gene from one organism for cross-fertilisation with another species. In this respect, genetic technology signals a movement from common methods of inter-species breeding to genetic cross-species breeding. Opponents of genetically modified organisms argue that genetically altered material may be transferred to other crops via insects, causing cross-pollination. As it happens, a series of problems with GM crops have already materialised in Britain. For instance, GM maize has damaged the wings of butterflies and modified oilseed rape has contaminated non-GM crops grown miles away from test sites. Flying in the face of public concern, many European governments have rejected the precautionary principle and elected to support the manufacture of GM foods. Despite conceding that genetically modified oilseed has already contaminated 1 per cent of the oilseed population, the British government have refused to limit the production of GMOs to enclosed laboratories, announcing plans to develop more outdoor testing sites.

Currently, biotechnology companies, ecological campaigners, politicians and food retailers are all vying to communicate particular 'stories' about GM foods to the public (Mintel Report, 1999: 3). Yet the manufacture of GMOs has provoked a surprisingly hostile reaction from the public (ESRC Report, 1999). Research studies indicate that GM foods are perceived as a significant health risk by members of the general public (see Cragg Ross Dawson Report, 2000; Mintel Report, 1999). It is quite possible that public anxiety about GM foods may be acting as a conduit for the expression of wider concerns about the relationship between humans and the environment (see Caygill, 2000: 155; CSEC Report, 1997: 3).

But what can we read into the story of GM foods? Is the disquiet over GMOs emblematic of the kinds of political conflicts which arise in the risk society? On the surface, public anxiety about GM foods lends weight to the claim that risk serves as a generator of political reflexivity. In Europe, a myriad of environmental, consumer and religious groups have vociferously countered the use of GM technology. In Austria, a referendum involving about a sixth of the population voted in favour of keeping Austria a GM-free zone (Adam, 2000a: 128). Meanwhile, in Britain, the idea of a five-year 'thawing-out' period without additional experimentation was supported by over 56 different non-governmental organisations. The ideological case of the anti-GM movement has gathered momentum, catapulted forward by lobbying groups such as Greenpeace. In a practical vein, GM food protesters have made use of direct political action, sabotaging GM testing sites in organised cells. As Adam writes:

> Europeans ... have responded with unusual strength of feeling to GM promoters' pronouncements that GM food is here to stay, that it is the future and that we had better get used to it. They are making their voices heard through opinion polls and demonstrations, by creating and joining anti-GM organizations, and by switching in large numbers to organically produced food. (Adam, 2000a: 129)

For Beck, the pressure exerted on dominant institutions to limit the production of GM foods is symptomatic of the wider 'subpoliticization' of society (Beck, 1998b: 16). For certain, the discord surrounding GMOs demonstrates that public opinion and micro-political actions can influence the behaviour of powerful companies (Williams, 1998). Public pressure has rewritten the business strategy manual, as illustrated by the rejection of GM foods by high-profile multinationals such as Unilever. In response to public pressure, there has been a steep general decline in the number of genetically modified products stocked by supermarkets.[8]

Recent research into public attitudes towards GM foods suggests that many lay actors possess a mature sense of the risks presented by genetic modification. In addition to the perceived risk to health, qualitative research indicates that the development of GM foods is construed as an inherently political issue. Reports commissioned by the Economic and Social Research Council (ESRC Report, 1999) and the Centre for the Study of Environmental Change (CSEC Report, 1997) describe a public culture of scepticism and suspicion. The development of genetically modified organisms is widely held to be driven by profit and power, not the public interest (see CSEC Report, 1997: 11). It would seem that the British government has made an unpopular and unilateral decision on genetic technology, choosing to adopt an unwaveringly pro-industry position (ESRC Report, 1999: 8). Tellingly, a sizeable number of respondents in the CSEC studies chose to draw no real distinction between the government itself and 'autonomous' regulatory bodies charged with overseeing the development of GMOs:

> The responses suggested that people have a general sense that they are not fully informed about food risks; that they tend to mistrust scientific claims of safety; that they question the motives of corporations involved in its development; and that they identify most with the voice of NGOs. (CSEC Report, 1997: 14)

These findings do key in with the risk society perspective, illustrating that critical reflection about the risks of genetic technology has nurtured a degree of public distrust in expert systems (CSEC Report, 1997: 11). This said, public attitudes toward GM foods are culturally specific and need to be properly contextualised. As Pidgeon (2000: 47) instructs, individuals do not approach risk issues as tabula rasa. In Britain, the current dispute around genetic modification is the latest in a sequence of food scares, including salmonella, listeria, e-coli and BSE. Accepting this historical trail, the strength of public feeling about food risk must be set against institutional mishandling of preceding incidents (see Cragg Ross Dawson Report, 2000).

RISK AS DISCOURSE

In the case of GM foods, it would appear that public pressure, allied to the direct political actions of NGOs, has successfully politicised the issue of food production and consumption. The conflict surrounding the development and distribution of GM foods acts as an exemplar of the subpolitical struggles which bespeckle the risk society. Nevertheless, we must avoid the temptation of committing to a theoretical position on the basis of fractional evidence. Again, Beck's fixation with the destructive aspects of risk militates against balanced presentation (Elliott, 2002: 312). Apparently risky situations can be read in different ways, depending on the inclination of the interpreter. To solidify this point, it is worth calling on the work of those who have elaborated the disciplinary effects of discourses of risk (Castel, 1991; Culpitt, 1999; Dean, 1999).

As discussed earlier, growing distrust in expert systems and an upsurge in subpolitical activity are characteristic features of the risk society narrative. Acceding to Beck's thesis, it is reasonable to suggest that the actions of institutional experts are increasingly questioned and monitored (Macnaghtan and Urry, 1998: 254; Prior et al., 2000: 111). Nevertheless, it remains debatable whether informal public scrutiny can be equated with a tangible movement towards a subpolitical culture. Opposing such a scenario, those from the governmentality school have identified a counter-trend toward the privatisation of politics through the employment of risk as discourse. Whereas Beck postulates that awareness of manufactured threats acts as a catalyst for political reflexivity, Foucauldians accentuate the disciplinary and restrictive functions of risk. Like risk, 'discourse' is an ill-defined term which has been subject to a range of usages within different disciplines. Although discourse has traditionally been linked to language, a broader definition of the concept has been developed in the social sciences. This wider understanding perceives discourses as sets of ideas, beliefs and practices that provide ways of representing knowledge. Discourse enables the presentation of certain forms of knowledge and precludes the construction of others (Woodward and Watt, 2000: 22).

But how does discourse relate to risk? As Lupton (1999a: 87) notes, information about risk has historically been collected by a range of experts, such as 'medical researchers, statisticians, sociologists, demographers, environmental scientists, legal practitioners, bankers and accountants, to name but a few'. Through this process, institutions have produced the language and data which form the basis for broader bodies of ideas. As social constructionists point out, in contemporary culture, medical, scientific and economic discourses govern what can and cannot be said about risk. Indeed, throughout western history, discourses have been utilised as a tool of mystification and social closure – for example, through the production of medical and scientific language which has excluded women and the working classes (see Douglas, 1985: 13; Woodward and Watt, 2000: 24). Foucauldians contend that, via the working of discourse we come to recognise and understand risk. Discourses of risk flow through the networks of social institutions which structure and govern everyday practice, making risk 'thinkable'. It is only through encounters within the family, education, media, the welfare state and employment that we come to 'know' about the existence and consequences of risks. Through the operation of discourse – as an idea and a material practice – what counts as knowledge about risk is determined. For Foucauldians, the debate is not geared around the health effects of risk, but how risks shape the way we experience our own realities: 'risk cannot be construed just as a potential threat to the self. Risk perception also involves the ways in which the self is able to perceive the self' (Culpitt, 1999: 23).

For those endorsing the governmentality approach, expert institutions employ discourses of risk to filter information, deflect opposition and reinforce dominant norms. Discourses regulate and discipline behaviour by generating 'truths' about society which become 'interiorised' by individuals (Foucault, 1978; 1980). The interiorisation of discourse enables people to make sense of the world and situates individuals in different 'subject positions' (Mackey, 1999: 127). Through this strategy, power relations are reproduced not by force or violence, but by discourses which facilitate patterns of self-regulation (Lupton, 1999b: 4). For Foucault, expert discourses of risk provide the parameters of appropriate action, serving as tools of regulation and surveillance (Caplan, 2000a: 23). Thus, institutional discourses are central to the construction of subjectivity, reproducing 'docile bodies' which do not threaten the political status quo. Flashing back to the risk society thesis, the discursive approach raises several disquieting questions about the effects of risk on political practices. Do discourses of risk flow through the capillaries of dominant institutions? Does the interiorisation of discourses of risk serve to dissipate political opposition?

According to Foucauldians, Beck underplays the operation of discourse and overlooks the possibility that risks can be marshalled to regulate everyday behaviour and stifle oppositional actions.[9] As Lupton (1999a: 8) notes, the rationality of risk is culturally ever-present, tumbling into an assortment of social fields. The process of childbirth becomes attached to a bundle of monitoring procedures, social workers provide estimations of harm, adverse weather and traffic warnings are issued and nutritional agencies impart information about food scares. Undoubtedly, a strong case can be made for the growing presence of institutional discourses of risk. However, the more crucial issue is the extent to which expert knowledge conditions risk attitudes and behaviours. In accord with the governmentality approach, prevailing institutions have historically used fears about risk to shape ideological discourses and to enhance the span of governance. The construction of the self is inevitably informed by institutionally generated knowledge and the circulation of expert discourses. Moreover, as cohesive discursive formations are formed, more intense forms of self-monitoring and regulation invoke patterns of social conformity which are difficult to resist (Lupton, 1999a: 88; Segal, 1997). As Schilling (1997: 65) observes, institutional warnings about health risks have encouraged people to 'keep' their bodies fit, healthy and active. As drafted in Chapter 6, in contemporary culture greater emphasis is placed on personal planning, particularly given the strain placed on welfare systems by an ageing population. In Britain, state welfare expenditure has been cut and the steady removal of welfare support has transported responsibility for risk toward the individual. At the very time when the discourse of risk is at its most advanced, governments appear to be strategically removing insurance systems and redistributing the burden of risk. In the British context, the changing language of welfare has echoed neo-liberal aspirations. The shift from welfare to privatised health and social security has been told in the tongue of the free market. Unemployed people have miraculously become 'jobseekers', citizens are transformed into 'consumers' and civil servants remade as 'welfare managers' (see Hughes and Fergusson, 2000: 142). To be just, Beck does recognise the ideological power of institutional constructions of risk. Nonetheless, the emancipatory bent of the risk society thesis dictates that discourses of risk are progressively carved open and challenged by subpolitical expressions. On the contrary, Foucauldians maintain that power is 'provocative', closing down rather than opening up political possibilities (Allen, 2000: 39): 'Discourses about risk are socially constructed narratives. Neo-liberalism constructed the discourses about welfare risk for its own hegemonic purposes ... in that sense neo-liberalism has used the anxiety about risk society for its own political ends' (Culpitt, 1999: 113).

A particularly prominent strategy of neo-liberalism has been the use of risk as a tool for political blame (Rose, 2000: 67). Butting against the risk society perspective, Foucauldians believe that the social journey of blame does not 'begin' from the risk and move outward toward the group. Rather, it begins with the group targeted for blame and affixes them to the risk: 'the significance of risk does not lie with the risk itself, but what risk gets attached to' (Dean, 1999: 131). Recognition of the ideological conferral of risk renders visible the bond between risk and social stigmatisation. The institutional attribution of responsibility for risk indicates that a distrust of 'otherness' can conveniently be forged into direction of blame (Lash, 2000: 51, Woodward, 1997: 15).

It would appear that the incongruity between the risk society perspective and the governmentality approach resides in their conflicting conceptions of power. Beck implies that political power has been exercised by institutions in a top-down fashion and needs to be replaced by a subpolitical bottom-up model. In contrast, followers of Foucault depict the relationship between risk and power as fuzzier and more dispersed. Theorists such as Culpitt (1999) have criticised Beck's objectivist construction of power, contending that power has never simply 'belonged' to dominant institutions, but courses through society as a whole. As a result, political power cannot simply be wrested back from institutions by individuals:

> It is not satisfactory to assume that discussions about risk can be tied solely to a revalorization of the pre-eminent power of individuals. In all of this it is the 'knowledge' of, and about, power that Foucault is attacking. Effective critique depends not so much on who has or does not have knowledge. It does not depend upon the sovereign/servant matrix. It is not about the power of inside knowledge *vis-a-vis* outside knowledge. It is rather about the structures and patterns of knowledge itself. (Culpitt, 1999: 42)

What is at stake then is a fundamental disagreement about the nature of power. Foucauldians favour a circular and networked notion of power. In opposition, Beck works with a laddered model of power which assumes that institutional discourses of risk are losing credence due to the opposition offered by new social movements. Whilst both theories perceive the individual to be self-monitoring, the concept of reflexivity allows Beck to attribute political agency to the individual. In contrast, in the Foucauldian version, techniques of self-surveillance produced by discourse are politically suppressive: 'What appears as the freedom of agency for the theory of reflexivity is just another means of control for Foucault, as the direct operation of power ... has been displaced by its mediated operation' (Lash, 1993: 20).

RECONFIGURING THE POLITICS OF RISK

Having critically reviewed risk society and governmentality projections of the political consequences of risk, it is time to delve deeper into the contemporary union between risk and politics. In comparison with Beck's image of the individual as reflexive actor, the Foucauldian perspective presents a more muted view of human agency. Unsurprisingly, parallels have been drawn between Foucault's theory of interiorisation and the Marxian notion of false consciousness. In both doctrines, individuals are seen as internalising institutional expectations that appear as reasonable and fair (Hughes and Fergusson, 2000: 37). Foucauldians maintain that the governmentality approach is more sophisticated than the theory of false consciousness, yet the idea of interiorisation still harbours residues of determinism. The problem with a pure Foucauldian approach is that it portrays individuals as insentient 'docile bodies', routinely complying with disciplinary discourses (Schilling, 1997). This may be the way the world works for some, but most people will critically reflect on their day-to-day activities and obligations. As contestation over genetically modified organisms illustrates, risk is a polythemic concept which produces diverse effects. Risks may, in certain circumstances, facilitate political opposition. In other situations, perceptions of risk may track the patterns of self-surveillance suggested by the discursive approach. Recognition of the profusion of strategies for dealing with risk delivers a decisive blow to the risk society thesis. Beck does acknowledge that individuals have become increasingly preoccupied with preventing and managing risk (1998: 12), but fails to enlarge on the corrective potential of risk. Institutional narratives of risk can serve to individualise blame and to constrain social solidarity (Furlong and Cartmel, 1997: 114). Historically, risk has been utilised as a tool of regulatory power, enabling governments to contain and deflect political opposition (Lupton, 1998: 88). For example, governmental discourses have sought to disown and individualise environmental risks, placing the burden of responsibility on citizens rather than the state. At the precise moment that consumers are exhorted to 'make the difference' by recycling, buying green products and conserving energy, the capitalist behemoth is extending its polluting production practices across the globe. In addition to fudging issues of environmental responsibility, the individualisation of blame can be utilised as a neat technique of concealment. Dominant institutions have the ability to exercise informational resources to ascribe danger to repressed groups (Scott, 2000: 40). In recent times, governmental discourses have blanketed asylum seekers, homosexuals and single 'parents' as risk-generating groups:

> For a person to be identified as posing a risk no longer means that she or he has to be individually observed for signs of dangerousness. It is enough that she or he is identified as a member of a 'risky population'. (Lupton, 1999a: 93)

As laid bare in Chapter 4, governmental discourses are often seized upon and amplified by the mass media, reproducing negative stereotypes. Through stereotyping people are steered toward the attachment of individual blame and away from critical reflection on the institutional reproduction of risk.[10] Stuart Hall et al.'s (1982) study reveals that labelling marginalised groups as dangerous 'others' can be an effective way of fostering compliance with restrictive law-and-order legislation. Thus, the social construction of discourses can serve to encourage the apportionment of blame, masking the multicausal reproduction of risk.[11] Arguing against Beck, in contemporary western cultures, the discourse of risk has enabled the state to pass through policies of risk regulation which uphold the dominant order. By reasserting social norms, the language of risk can be used to fortify institutional control and reinforce unequal relations (Lupton, 1993: 431). Risks can, and indeed *do*, stimulate conservative responses which restrain autonomy and creativity (Caplan, 2000a: 23; Furedi, 1997). The end game of advocating over-zealous avoidance strategies is evocatively played out by Castel:

> A vast hygienist utopia plays on the alternate registers of fear and security, inducing a delirium of rationality, an absolute reign of calculative reason and a no less absolute prerogative of its agents, planners and technocrats, administrators of happiness for a life to which nothing happens. (Castel, 1991: 289)

Although Castel's case is overstated, there is a strong connection between discourses of risk and the ordering of human behaviour. Beck recognises this affiliation in industrial society, but sees the risk society's regenerative subpolitics as the death-knell for institutional expertise. However, it cannot be super-induced that risk awareness will force the general public to attribute liability to expert systems, less still become political agitants. Turning these soils, several theorists have been circumspect about the emancipatory possibilities of a 'politics of risk' (Abbinnett, 2000; Nugent, 2000; Rustin, 1994). Beck is justified in identifying a widening trend of activity outside the formal democratic system, but the transformatory power of subpolitical activities remains to be seen. As indicated in Chapter 3, it would be a mistake to arbitrarily place NGOs on a political pedestal. Although campaign groups such as Greenpeace have attempted to engage at a local level, they remain firmly bound to the lifestyle concerns of the First World:

> Greenpeace's environmentalism shares some of the globalist assumptions and practices of its enemies. It uses 'big science' to define environmental issues in global terms rather than in local, culturally sensitive contexts ... and it maintains a freewheeling distanciated relationship from particular localities. (Tomlinson, 1999: 191)

What is more, it is unlikely that the bulk of global political activities – be they inside or outside of the formal process – are principally driven by risk. On a global scale, current tensions between the East and the West are the result of an admixture of colonialism, neo-imperialism, contrasting religious beliefs, discordant political values and unequal resource distribution. Locally, intra-national struggles within Spain, Canada and Ireland are historically rooted in matters of cultural identity, rather than risk. Moreover, notable subpolitical expressions – such as the recurrent anti-globalisation protests – are about much more than risk. It is feasible that direct political actions are more the result of general disenchantment with the functioning of liberal democracy, than a reflexive response to unsafe and uncertain social conditions.

It must also be noted that subpolitical activities cannot be detached from the formal process. Whilst Beck sees subpolitics as an external process of political regeneration, subpolitical groups are susceptible to engulfment by the formal process. In recent times, national governments have attempted to suck non-governmental organisa-tions in to the existing political system via consultation and round-table discussion. This trend is particularly prevalent in risk-related areas such as food safety, the environment, crime and drugs. The capacity of national governments to defuse political opposition must not be discounted. Unfortunately, Beck's approach to the politics of risk plays down the reflexivity of national state institutions in reconfiguring political structures. It must be remembered that the globalisation of politics has been put in train by traditional political forces as well as subpolitical groups. In response to the emergence of transboundary issues, there has been a distinct internationalisation of formal politics and a dramatic rise in the number of intergovern-mental panels and agencies.[12] Again, the ability of national governments to dissipate subpolitical power by incorporating NGOs into the system and generating global political structures is disregarded in the risk society perspective.

Beck also fails to appreciate the ability of the state to coerce and disempower subpolitical protests. The British state has a long tradition of aggressively dealing with oppositional activity, as evidenced in the recent policing of anti-capitalist and anti-war demonstrations. Protesters campaigning against the policies of the World Trade Organisation in Italy, America and the Czech Republic have been greeted with equally repressive measures. In modern times – as in previous eras – the coercive potential of the state can serve to dissuade individuals from involvement in oppositional politics (Hillyard and Percy-Smith, 1988; Lodziak, 1995). As the modes of direct political action evolve, so too do methods of state control. At the 2003 G8 summit in Evian, over 26,000 French and Swiss soldiers and police were mobilised to counter anti-globalisation protesters (Lichfield, 2003: 2). In order to defuse the power of direct actions, a 30-mile exclusion zone was imposed, preventing demonstrators from accessing the location of the meeting.

Beck's theory of subpolitics does not attend to the coercive capabilities of particular nation states and seems to take place in an abstract global space. The theory of subpolitics foresees a veritable political reformation, with new structures, institutions and debating mechanisms flowing from active public debate. At a macro level, an 'upper house of technology' would be introduced to regulate large-scale political decisions. On the ground, subpolitical groups could debate key political issues within a deliberative democratic framework (Beck, 1995b: 506). Such an idyllic scenario invites us to steal Bauman's (1992: 217) waspish depiction of the Habermasian ideal: 'society shaped after the pattern of a sociology seminar, that is, there are only participants and the one thing that matters is the power of argument'. At a time when postmodern theorists have been critical of universal truths and grand narratives, Beck remains unashamedly committed to the goal of rational consensus (Beck, 1998: 21; Rustin, 1994: 394). Nonetheless, as well as being *for* the people, democracy has to be established *by* the people. It should not be assumed that people are inherently inclined towards radical reformation of democratic procedures, nor that they will be universally welcoming of the extra burden of political responsibilities. A meaningful movement toward post-industrial socialism would require a pendulum shift in consumption patterns, cultural practices and social values. For some, the jump from ideological approval to practical application may prove exigent: 'Those espousing a "Third Way" which actually takes on an accurate global view should be preparing their constituents to accept rather grim costs: no winter shoes for the kids this year, or next' (Nugent, 2000: 232).

Underpinning the theory of subpolitics is the idea that enhanced forms of democracy may deliver a net reduction in the production of risks. Using the example of transport policy in Munich, Hajer and Kesselring (1999) take issue with this assumption, demonstrating that the development of democratic structures does not ensure the elimination of environmental risks or the development of subaltern discourses. In the Munich case, a variety of deliberative methods such as referendums and round table discussions were used to improve the quality of the decision-making process. However, the assimilation of deliberative practices failed to produce a reduction in environmental pollution. Lamentably, it is likely that many of the democratic alternatives envisaged by Beck have already been co-opted by political and economic power brokers (see Hajer and Kesselring, 1999: 14; Kerr and Cunningham-Burley, 2000: 293). In the economic marketplace, commercial gurus have become increasingly adept in the art of 'greenwash', a technique of reverse spin through which risk-producing companies are represented as environmental champions (Matthiessen, 1999). Forward-thinking risk-sensitive companies have held public consultations and open meetings in a bid to cancel out negative publicity. To date, attempts to involve subpolitical groups in the political decision-making process have involved limited forms of participation and have done little to challenge existing power relations:

By themselves ... 'consultations' and 'participation' do not necessarily solve the problems ... Who is consulted? Who participates? Who decides who is consulted and who participates? Who decides what the issues are that people shall be consulted on? What counts as relevant knowledge and expertise? Is anyone obligated to pay attention to the consultation, or is the simple process of staging a consultation considered sufficient? Unless these questions are discussed, 'consultation' and 'participation' are likely to prove merely new ways of containing – or even silencing – popular environmental concerns. (Purdue, 1995: 170)

As Purdue (1995: 171) notes, the appropriation of supposedly democratic methods can afford risk producers 'public consent' via the gloss of partial forms of consultation. Hence, the parallels with Foucauldian discourse theory again arise, with the possibility that public consultation forms part of a wider culture of control:

> The motive for the introduction of new participatory practices is just as likely to be about enhancing the effectiveness or institutional capacity of government as it is about a democratisation of policy making. When it comes to the assessment of what the role and function of participatory practices in a risk society actually requires, we need to carefully consider the way in which the new practices of governance relate to one another. (Hajer and Kesselring, 1999: 19)

As Hajer and Kesselring concede, deliberative democratic practices may in principle offer a way forward for more interactive forms of political decision making. However, democratic practices are just as easily tagged on to existing political procedures, attenuating rather than enhancing institutional accountability for risk.

To do justice to the risk society perspective, Beck recognises that, in isolation, subpolitical activity cannot act as a panacea (Beck, 1999: 131). In *The Reinvention of Politics* (1997: 5) alternative political routes into 'counter-modernisation' – such as nationalism, political violence and scapegoating of 'enemy stereotypes' – are unloaded.[13] Nonetheless, the negative aspects of subpolitics are talked over and treated as an addendum. Sufficient emphasis is not afforded to the negative uses of direct political actions, particularly in the light of 9/11, new forms of bioterrorism and the continuing conflict in Palestine. Lamentably, the tendency to treat politics as a corollary of risk adversely affects the balance of the risk society argument. In a roundabout way, this brings us full circle to the issue of differentiated risk perception discussed in Chapter 5. In the first instance, forms of stratification and cultural identity will affect the formulation of political understandings of risk. Individuals will possess a range of cognitive perspectives on risk which mutate according to social roles and positions (Mythen et al., 2000: 16). As such, political attitudes can be expected to evolve over time, space and place, with uniform opposition to risk being atypical. Remembering the hawkish criticisms raised by Nugent, it would be prudent to remain alert to possible inconsistencies between expressed opinions, cultural attitudes and actual behaviour towards risk. Political perceptions of risk are likely to be fluid and ambiguous rather than absolute. For example, in the midst of a general rejection of GM foods, Grove-White and colleagues also detected a fatalistic attitude towards GM foods, with many viewing the continued diffusion of genetically altered foods as inevitable (CSEC Report, 1997: 1). Not only does this caveat chime with the governmentality approach, it also reminds us of the break-beat between risk consciousness and political action. In order for subpolitics to flourish, the 'value–action' gap between risk consciousness and political activity needs to be bridged. Although risk research has turned up little active endorsement of existing political institutions, responses to the diffusion of GMOs illustrate that feelings of powerlessness remain commonplace:

> The development of genetically modified foods appeared to be seen as lying outside people's control, with little sphere for public choice or intervention ... these feelings of inevitability seemed to reflect a felt absence of choice and a sense that, realistically speaking, the technology was unstoppable. Such inevitability appeared to lie behind feelings of passive resignation in the majority of the groups. (CSEC Report, 1997: 13)

These are significant findings as far as the relationship between risk and political mobilisation is concerned. Insofar as Beck sees oppositional activity being organically generated by risk, the evidence demonstrates that political reflexivity will be just one response amongst many. As discussed in Chapter 2, despite general distrust of expert systems, empirical studies have not reported rejection of institutional structures per se. Rather than outright dismissal of risk-regulating structures, what emerges instead is a more tempered opinion that existing institutions are functioning ineffectively. This is backed up by the expectation that existing regulatory bodies will continue to monitor risk in the future (CSEC Report, 1997). Thus, there remains a strong residual expectation that expert institutions will adapt monitoring procedures to deal with the challenge of risk (ESRC Report, 1999). This is indicative of a distinct cognitive ambivalence which does not tally with Beck's assumptions of radical political change.

So, where do these findings leave us in relation to our two overarching approaches towards the politics of risk? Ironically, the criticisms directed at Foucault's understanding of the individual can be turned against Beck's theory of reflexivity. The risk society subject is also something of an empty vessel; albeit one waiting to be activated by risk, rather than inscribed by discourse (Lash and Urry, 1993: 32). Unfortunately, both the risk society thesis and the governmentality perspective operate at the lofty level of grand theory and tend to assume a universal 'risk subject' (Lupton, 1999b: 6). This mutual problem arises out of an impracticable desire to uniformly foretell the political effects of risk. Both the governmentality and the risk society perspective suggest predictable – if contrary – political outcomes to risk situations. Where discourse theory overplays the totality of social structure and dims individual agency, Beck underplays the coercive capacity of social structures and exaggerates individual agency. Thus, whilst we can use discourse theory to bring up the limitations of the theory of political reflexivity, the risk society accent on human autonomy flags up the shortcomings of the Foucauldian approach.

CONCLUSION

As we have moved through the chapter, it has become apparent that the relationship between risk and politics is complicated and barbed. As far as the extent and emancipatory potential of subpolitics is concerned, the evidence is inconclusive. In support of Beck, there does appear to be general disenchantment with the current political system and a broader trend of scepticism toward expert systems. To boot, public attitudes towards the introduction of genetically modified foods indicate that embryonic forms of political reflexivity are growing around risk issues. And yet risk conflicts do not robotically produce 'coalitions of opposites' and/or mutually agreeable political outcomes. As illustrated by the GM case, risk conflicts can easily result in value entrenchment, as parties with irreconcilable perspectives down tools and dig in (Sparks, 2003: 202). Contra Beck, discontinuity remains between reflexive engagement with risk information and political mobilisation. Public criticisms of expert systems must not be read off as an inclination to radically transform socio-political structures. As Eagleton (2003: 82) prudently reminds us, 'it is rational to resist major political change as long as a system is still able to afford you some gratification, however meagre, and so long as the alternatives to it remain perilous and obscure'.

Taking on board the criticisms raised by Foucauldians, we must recognise that expert discourses can serve to individualise coping strategies, promote the unwarranted attachment of political blame and intensify strategies of surveillance. Beck's desire to attribute political reflexivity to the individual glosses over the possibility that the language of risk may reinforce as well as undermine social control. Whilst the Foucauldian critique has enriched the general debate about risk, the discursive approach has itself been rightly criticised for presenting a passive and disembodied subject. As Connell (1995: 56) argues, discourse theory tends to depict individuals as ahistoric, blank slates on which disciplinary power is written. Along with Beck, Foucauldians have failed to grasp the manner in which stratification and cultural identities influence individual understandings of risk. On a broader note, these criticisms indicate that abstract theorising about the politics of risk has led social theory into something of an impasse. Foucauldians have maintained a position as relativist as the risk society perspective is realist. Consequently, the two sides have simply talked past one another (Lupton, 1999b: 6). To inject momentum into the debate, greater recognition of the diversity of the politics of risk is required. Whereas risk society and governmentality perspectives provide valuable contributions to the debate, neither approach adequately captures the inherent untidiness of the dense relationship between risk and politics.

Conclusion

In applying the risk society theory to contemporary cultural practices, both productive and unconstructive dimensions of the argument have come into view. By way of conclusion, I wish to restate the headline findings and reflect on the significance of Beck's venture for the way in which risk is understood, both in academia and wider society. To this end, it is worth rounding up the major criticisms made in the book and offsetting these against the political value of the risk society thesis.

On the negative side, the risk society argument is plagued by both theoretical and empirical deficiencies. Beck's determination to provide a universal model of risk helps us to understand his general unwillingness to engage in the process of empirical validation. Naturally, the decision to prioritise theoretical sanctity over empirical detail has been a source of much ire and consternation (Alexander and Smith, 1996; Dryzek, 1995; Hajer and Kesselring, 1999: 3; Marshall, 1999). It is certainly true that the conceptual framework of the risk society produces an excessive amount of stickiness and this makes empirical calibration difficult. For some, Beck's 'sociological spoon bending' has been decoded as an attempt to conceal the contradictions inherent within the risk society thesis (Smith et al., 1997: 170). At the very least, the theoretical inconsistencies contained within Beck's argument give rise to a lack of constancy. Over the course of the book, I have identified seven key areas of theoretical weakness which bog the risk society perspective down.

First, it has been demonstrated that the attempt to collapse variant forms of danger into bipolar categories clusters together disparate threats and overstates epochal differences (Anderson, 1997: 188). The inherent diversity of risk militates against a crude division between natural hazards and manufactured risks. Whilst the nature of risk has altered over time, by exaggerating the margins between epochal dangers, Beck reproduces an unfeasibly tidy historical narrative (Alexander and Smith, 1996; Strydom, 1999: 53).[1] The alleged phases of modernity are too monolithic to bear resemblance to the diversity of cultural experiences:

It is clear that the genealogy of risk is much more complex than the theory of risk society allows. Risk and its techniques are plural and heterogeneous and its significance cannot be exhausted by a narrative shift from quantitative calculation of risk to the global-ization of incalculable risks. (Dean, 1999: 145)

Second, the notional attempt to equalise patterns of distribution is exposed by the recurrent flow of risk through the locks of class, gender, age and ethnicity. Despite anomalous instances of boomerang effects, the dispersal of risk invariably reinforces rather than transforms existing patterns of social inequality. The failure to properly acknowledge continuities in social reproduction can be traced back to a third theoretical short circuit, emerging out of a reliance on worst imaginable accidents as the archetypal form of risk (Scott, 2000). Beck's concentration on hypothetical 'icons of destruction' as opposed to risks which routinely impact leads him to extrapolate too readily from worst-case scenarios and to overstate the globalising tendency of risk. As a consequence of this slippage, the uneven diffusion and differential impacts of risks are concealed (Engel and Strasser 1998: 94; Hinchcliffe, 2000).

Fourth, having synthetically fused the effects of risk, Beck is obliged to maintain that public perceptions of danger will be uniformly ordered. In opposition to the risk society perspective, cultural under-standings of risk cannot be generalised and will be diverse, rather than homogeneous. Public attitudes towards risk are multifaceted and will be proselytised through established networks of families, friends and work colleagues (Reilly, 1999; Tulloch and Lupton, 2003). People do not share the same life experiences. Ergo, they cannot possibly share the same interpretations of risk. Fifth, the risk society perspective is inhibited by its unconditionally negative conception of risk as harm. It should be remembered that risk is understood as a phenomenon that promises beneficial as well as detrimental consequences (Lupton, 1999a: 148). While Beck paints a dystopic picture of risk, we must not forget that risk taking can be a socially progressive process (see Giddens, 1999: 2; Lupton and Tulloch, 2002a). In economic language, there are 'positive risks' from which one can only gain, 'neutral risks' where one can lose or gain and 'negative risks' where one can only lose. Beck is hooked on the latter category and this effectively reduces risk to lose-lose situations. On reflection, it is clear that risk taking has led to immense technological, medical and economic advancements. For the philosophical, risks may be rationalised as the Faustian bargain for the benefits of modernisation (Irwin et al., 2000: 95). Unfortunately, the risk society thesis is purblind to risks that are independently taken by individuals and ignores the salutary role of risk taking in social development (Culpitt, 1999: 113). As Castel (1991: 289) argues, we perhaps need to strike a balance between our fears about catastrophic risks and a life in which very little happens.

Sixth, the unbroken use of a restricted version of risk encourages the view that human beings are innately risk-averse. In reading *Risk Society* (1992) one gets the impression that nothing short of a life of enduring security will suffice. From an existential pitch, we might question the value of a culture that attempts to disconnect itself from danger. In western society, subcultural groups may seek to mobilise risk as a tool for flouting convention and challenging authority (Lupton, 1999a: 167; Lyng, 1990). At the margins, the heterogeneity of social meanings means that one person's risk may constitute another person's pleasure. Whilst the risk society thesis is bound to index risk to harm, in reality the concept of risk is 'as long as a piece of string and as elastic as bungee rope' (Eldridge, 1999: 106).

Finally, if we stretch the risk society over the global geographic, yet more bumps and cracks appear. Within non-western cultures, risk taking may be used as a mechanism for performativity and a source of social cohesion (Douglas, 1985: 26). Taking a global perspective, risk should be understood as a polyseme, not an essential and immutable category (Caplan, 2000a: 18). Given that local experiences are cultivated and situated, we must be aware that the meaning of risk will always be fixed in the eye of the beholder (Fox, 1999: 13). Beyond this, anthropologists have noted that Beck's totalising approach to risk is tainted by a distinctly Eurocentric bias (Bujra, 2000: 63; Nugent, 2000: 236). The risk society metanarrative of modernisation may unwittingly reproduce an evolutionist and westernised model of social development. The combined weight of these shortcomings suggests that the underlying theoretical assumptions of the risk society perspective are unsound.

On a more positive note, in applying the risk society thesis to grounded research, support has been registered in several areas. For example, empirical evidence broadly endorses a relative rise in public risk consciousness, perceived unmanageability of manufactured risks and the increasing individualisation of cultural experience. Since the publication of *Risk Society* (1992), many western cultures have witnessed heightened institutional scrutiny, shifts in the relationship between experts and the public and more cohesive forms of subpolitical activity. While the risk society thesis is empirically light, simply highlighting Beck's disinclination for hard data does not disprove the formation of social trends: 'absence of evidence is not the same thing as evidence of absence' (ESRC Report, 1999: 7). In effect, several of Beck's detractors may have errantly thrown out the empirical baby along with the theoretical bathwater.

The melodramatic reactions which the risk society thesis has provoked amongst some critics may stem from a failure to comprehend the broader objectives of the risk society narrative (see North, 1997; Smith et al., 1997). Although Beck should not be rendered immune to criticism, in many respects his project does not compute with academic tradition. The heady mix of impassioned critique, paradoxical prose and provocative irony have made *Risk Society* (1992) an easy target for academic purists. Doubtless, Beck's popularity has not been enhanced by his simmering criticisms of the sociological tradition (see Beck, 1997: 17). The empirical precision esteemed within the social sciences does not sit comfortably with Beck's method or style of narration. The trajectory of the risk society thesis departs from sociological tradition in a number of ways. First, the dark wit and humour present in Beck's writing has more in common with works of popular fiction than scholarly texts (Goldblatt, 1995: 154). Indeed, in his native Germany, Beck is considered to be a storyteller as much as a serious social scientist (Lash and Wynne, 1992: 1).[2] The risk society thesis is assembled in the spirit of exploration; it is motivated not by assiduous empiricism, but by challenging the sociological canon and creating stimulating ways of thinking through the modern condition. To enjoy the visionary quality of Beck's work, the normal academic rules of engagement need to be temporarily suspended:

> For all its problems, the work of Ulrich Beck retains an electric quality. Idea after idea jumps off the pages of his work. Some lack precision, others never receive justification, and still others contradict one another. Qualifications sit on top of one another; arguments disappear only to appear once again; fuzzy slogans compete with the claims of common sense. But then come the golden nuggets of dazzling insight. (Bronner, 1995: 85)

As we have seen, Beck's sociology does not provide us with a daintily penned portrait of contemporary society. Rather, it yields a blurred snapshot of the interchange between the present and the future. Indeed, Beck willingly admits that his work paints the changing shape of the world in 'broad-brush strokes':

> To put it bluntly, I am perhaps the least certain participant in the uncertain stance in which I deal. The lack of ifs and buts in the formulations is a question of style. Let this fact be taken out of parentheses and writ large once and for all. (Beck, 1995a: 13)

Further, many critics have been insensitive to the historical and contextual factors which have shaped the risk society project. Beck's metatheoretical approach is incontrovertibly rooted in the German sociological tradition and needs to be positioned along a continuum which includes Marx, Weber and Adorno (Lash and Wynne, 1992: 2).[3] Like his predecessors, Beck does not succeed in providing an exhaustive and watertight sociological theory. To relate the broader dimensions, overarching models are always prone to magnification. Hence, the risk society perspective is best treated as an heuristic device which allows us to observe and probe the peculiarities and perils of modern life. Of course, the risk society thesis is littered with faults, but these faults have generated the very dialogue through which academic and social knowledge has been advanced.[4] Characteristically, Beck (2000d; 2000e) has warmly welcomed the argumentation that has feasted off his work. Within academia, Beck's oeuvre has provided a bridge between previously detached disciplines. Long overdue dialogue about risk has elucidated common and conflictual ground between subject areas previously incommunicado.

We must also be mindful of the fact that the risk society thesis is at once a political and an academic project (Beck, 1995a: 12; 1997: 5). In many ways, *Risk Society* (1992) acts as a clarion call for the radicalisation of modernity: an entreaty for the reformation of politics, the economy and science. Progressive utopian demands are not always consonant with the rigorous requirements of academic theory building. To his credit, Beck (1992; 2000b) recognises the unrefined nature of the risk society perspective. It is almost as if he, like Baron Frankenstein, is obliged to nurse the monster, despite recognising its unruliness. In this sense, empirical and theoretical accuracy may be the trade-off granted for political effect:

> Believed risks are the whip used to keep the present day moving along at a gallop. The more threatening the shadows that fall on

the present day from a terrible future looming in the distance, the more compelling the shock that can be provoked by dramatizing risk today. (Beck, 2000d: 214)

As this quote illustrates, the risk society thesis is designed to unsettle the collective conscience, to force us to confront our social demons. Living in the risk society means facing up to the porosity of boundaries between nature and culture, local and global, public and private (Adam and van Loon, 2000: 5). Outside academia, productive media and public discussion about risk has acted as a vent for the expression of ethical and moral concerns. Well publicised cases of institutional incompetence have undermined public trust in experts and acted as a stimulus for political contestation. Beyond the routine spin and fudge of party politics, audible subaltern voices have challenged the objectivity of institutional methods of risk assessment. Risk practitioners are now realise that they must do more than simply 'get the numbers right' (Fischhoff, 1995). Risk-regulating institutions have recognised the need to be sensitive to public opinion and to engage with the aspirations of diverse stakeholder groups (Dean, 1999: 144; Handmer, 1995: 91; Mythen, 2002). It remains to be seen whether these developments signal a deeper commitment to openness, or if they are merely part of a dexterously constructed facade.

What is certain is that the emergence of irremediable risks does not guarantee a one-way journey towards political emancipation (Tomlinson, 1999: 206). Risks are not routinely ignited in the public sphere leading to rational argumentation between expert systems and subpolitical groups (Smith et al., 1997: 171). Given the diverse interests and viewpoints of non-governmental groups, it cannot be assumed that subpolitical activity will lead to the formation of an emancipatory politics (McGrew, 2000: 146). Despite Beck's tag as a 'prophet of hyper-enlightenment' (Szerszynski et al., 1996), we would do well to remember that, in matters of risk, the truth is not always 'out there'. A meaningful political dialogue about risk must be sensitive to issues of contingency and cultural difference. There is no 'right' way to define, negotiate or regulate risk. The meaning of risk will be infinitely contested, and reasonably so (Wilkinson 2001: 99). The most fertile route forward involves propagating a locally sensitive social dialogue about the new types of risk and uncertainty which characterise the modern age. However, as the fault lines of risk expand, we must not lose sight of society's oldest burden, the crushing weight of poverty. The question of how a presently segmented politics of risk can be incorporated into a materially effective and unified emancipatory project is yet to be determined. At present, lack of access to power-bound spaces has left new social movements with the pressing concern of how to convert subpolitical energies into a vibrant and functional democratic system.

In conclusion, for all its faults, we need to hold on to the inferences of the risk society thesis for social policy, politics and public life. Significant historical landmarks such as 9/11, the BSE crisis and the manufacture of biological and chemical weaponry warn against blithely plodding on with outmoded strategies of risk management. As global risks continue to evade the national structures of modernity, the resonance of the risk society thesis is reaffirmed. Of course, this does not detract from the academic task of deconstructing and repairing the risk society model. On the contrary, to retain explanatory potential, macro-structural perspectives must be complemented by grounded micro-level research. In interrogating, disassembling and rebuilding Beck's work we are formulating searching questions about the world we live in, and, moreover, the one which we are destined to inhabit.

Notes

INTRODUCTION

1. Extract from Tony Blair's address to the US Congress, Friday 18 July 2003. The full speech can be seen at <www.politics.guardian.co.uk/labour/story0,9061,10000564,00.html>
2. It is estimated that for every death from disease today there were over a hundred in the Middle Ages. Infant mortality rates have also declined steeply. In Britain in the year 2000, 6 children per 1,000 died before the age of one, as compared to 18 in 1981 and 84 in 1921 (Social Trends, 2002).
3. Beck (1998b: 12) and Giddens (1994) also use the term 'manufactured uncertainties'. To avoid confusion, I will stick to 'manufactured risks' throughout the book.
4. The book was first published in Germany in 1986 under the title *Risikogesellschaft: Auf Dem Weg in ein andere Moderne*. Henceforth, I will abbreviate the title to *Risk Society* (1992).
5. It should be emphasised that these twin objectives course through each chapter of the book, rather than being approached consecutively.
6. In Chapter 5, I draw upon the anthropological and psychometric perspectives to enrich Beck's understanding of risk perception. In Chapter 8, the governmentality approach is mobilised to illuminate the relationship between risk, discourse and politics.
7. A number of significant concepts formulated in *Risk Society* (1992) are developed and refined in *Ecological Politics in an Age of Risk* (1995a) and *World Risk Society* (1999). This said, there are also notable differences in the style and content of each book. *Risk Society* (1992) is comparatively broad-based, reviewing changes in employment, gender and social relationships in relation to risk and individualisation. *Ecological Politics in an Age of Risk* (1995a) has a narrower base, being more acutely focused on the impacts of environmental risks on the natural environment. Meanwhile, *World Risk Society* (1999) accentuates the increasingly global effects of techno-scientific risks and matches these to transformations in political participation.
8. See, for example, the use of multiple tenses in Chapter 7 of *Ecological Politics in an Age of Risk* (1995a).

CHAPTER 1

1. The book has been translated into more than 15 different languages, including Japanese, Russian and Chinese.
2. For a rich discussion of the changing meaning of risk, see Bernstein (1996), Boyne (2003) or Lupton (1999a).

3. As will be discussed in Chapter 5, Beck perceives risk through a variety of different lenses. In *Risk Society* (1992) and *World Risk Society* (1999) Beck oscillates between realist and social constructionist understandings of risk. On this point, see Alexander and Smith (1996) and Lupton (1999a: 59–62).

4. For further explanation of these epochal transformations, see Beck (1995a: 78).

5. Nearly 750,000 people are thought to have died as a result of the Chernobyl disaster. Further, the incidence of thyroid cancer in Belarus and the Ukraine has risen rapidly in recent years. For further details about the human consequences of the explosion see <www.chernobyl.org.uk>

6. Throughout his work, Beck refers to a range of epochal distinctions, including: pre-modernity; pre-industrial high cultures; simple industrial modernity, classical industrial society, industrial risk society, reflexive modernisation, residual risk society and the risk society. The grounds for these distinctions are most coherently explained in *Ecological Politics in an Age of Risk* (1995a: 78). A digestible review is provided by Goldblatt (1995: 167).

7. The interregnum between industrial society and the risk society proper is referred to by Beck as 'industrial risk society'.

8. As will be rendered explicit in Chapter 5, the risk society thesis assumes homogeneous perceptions of danger within and between historical eras.

9. In Britain, the Labour Party claimed a landslide victory in the 1997 General Election with a manifesto based around modernisation of key social institutions and the generation of a 'stakeholder society'. Despite such political aspirations, the gap between the richest and the poorest classes in Britain has continued to expand under Labour's tenure (see Mackintosh and Mooney, 2000). Indeed, several Labour ministers have publicly admitted that the party has failed to significantly reduce disparities in income, education and health (see Grice, 2003: 4).

10. For a concise comparison of the work of Beck and Giddens, see Lupton (1999a: 58–83).

CHAPTER 2

1. The impacts of individualisation on everyday life will be considered in Chapter 6 and the political effects of a shift in social distribution will be assessed in Chapter 8.

2. World energy consumption is thought to have doubled in the 20 years between 1973 and 1993 (see Kamppinen and Wilenius, 2001: 312).

3. Greenhouse gases include carbon dioxide, methane, nitrous oxides and chlorofluorocarbons.

4. Of course, the real problem is not so much that the planet is heating up per se, more that human activities and practices are causing it to do so.

5. Statistic cited in *Society Matters* 5, Open University Bulletin: 12.

6. Drawn from *Society Matters* 5, Open University Bulletin: 12.

7. Figures taken from 'The Information', *Independent,* Saturday 24 May 2003: 58.

8. Greenpeace currently has over 2.5 million members in 158 countries across the globe.

9. We would do well to remember that 40 per cent of carbon monoxide fumes and a fifth of total greenhouse gas emissions are produced by motor vehicles.

10. As Macnaghtan (2003: 69) suggests, in recent years there has been a feint movement away from collective public understandings of the environment ('out there') to a more privatised understanding of the local effects of global changes ('in here').

11. One in five adults in the UK subscribes to a non-governmental environmental organisation of some kind (ONS, 2001).

12. Given that the US consumes over a third of the world's energy, nurturing environmental consciousness in America ranks as a higher priority than in relatively Green countries, such as Belgium or the Netherlands. With the US administration having refusing to accord with a string of environmental treaties in recent years, the portents are not favourable.

CHAPTER 3

1. A more acute examination of the shifting relationship between institutional 'experts' and 'lay actors' will follow. In Chapter 7 the contracting space between experts and lay actors is explored. In Chapter 8 we consider the implications of recent modifications in the form and content of politics for expert–public relations.

2. As will be explained in Chapter 4, Beck casts the mass media as a reluctant partner in the relations of definition.

3. For a classic exposition of scientific rationality see *The Royal Society Report on the Public Understanding of Science* (1986). An outstanding critical commentary is provided by Brian Wynne (1989: 34–6).

4. This said, we should be wary of imprecisely applying particular empirical studies to the risk society thesis. For example, the surveys employed by Dickens were not designed to directly probe lay–expert conflicts, apropos issues of risk. Whilst carrying undoubted weight as a yardstick of public opinion about science, quantitative surveys do not have the capacity to provide us with explicit knowledge about the lay–expert divide, as articulated through specific discourses of risk.

5. I will elaborate on the relationship between trust and reflexivity in Chapter 7.

6. In relation to risk distribution, most would agree that the Swedish welfare system has been more effective than the Romanian model in providing material security for its citizens. In terms of risk perception, public sensitivity to the potential dangers presented by GMOs is more acute in Britain than America.

7. For example, it is legally acceptable to consume cannabis in designated areas in Holland. In contrast, in Croatia even wearing an item of clothing which carries the word 'cannabis' is a criminal offence.

CHAPTER 4

1. The generality of the term is problematic. The 'media' describes a host of outlets and portals including television, radio, newspapers, magazines and the internet. Unfortunately, in this chapter we do not have adequate space to differentiate between media forms in any detail.
2. Mimicking the conceptualisation of the media in the risk society thesis, we will primarily focus on mainstream broadcast journalism.
3. In the wake of 9/11, the threat of bioterrorism has emerged as an issue which has consumed much airtime in the United States, generating anxious questions about the sturdiness of national security measures.
4. In Britain, public-service broadcasting is the exception to this rule. However, public-service outlets cannot be insulated against a competitive market and remain driven by audience ratings. This was recently made explicit when BBC News put back its regular 9p.m. time slot in an attempt to poach viewers of the later ITN Nightly News.
5. It is worth noting that these are the very institutions through which public knowledge is constructed in the risk society thesis.
6. Routine publication of press releases is common practice within government departments such as the Department of Health and the Department of Transport, Environment and the Regions.
7. Mainstream media reporting of the 2003 invasion of Iraq stands as a case in point. Prior to Allied aggression, the British broadcast media devoted a sizeable amount of coverage to public opposition to the conflict. However, once the war commenced this critical strand of reporting all but disappeared.
8. The use of photographic imagery has been found to have important impacts on the meaning made of risk events (see Jones et al., 1997).
9. *Sun* editorial column, 17 November 1989. As cited in Eldridge (1999: 113).
10. As a BBC environmental correspondent interviewed by Anderson (1997: 121) states: 'we're about pictures ... we're about words as well, but words are captions to pictures, essentially.'
11. A number of widely watched films have explored the theme of risk, including *Threads, Blade Runner, Safe, Jurassic Park* and *Crash*.
12. Via cross-cultural content analysis, Hansen discovered that reports on nuclear energy were more prevalent in English than Danish news bulletins. One possible reason for this disparity is that Denmark is bereft of nuclear industry, whilst nuclear power is still an important source of employment and wealth in Britain.
13. This phase of Reilly's longitudinal research was undertaken after the first wave of media concern about BSE in 1992. As the seriousness of the situation became apparent, later research in 1996 indicated much higher levels of public concern.
14. This is particularly pertinent to risks which emerge post hoc, where damage to personal health may already have occurred.

CHAPTER 5

1. For analytical purposes, the objectivism/realism and relativism/construc-tionism couplets can be used interchangeably (Lupton, 1999a: 35). Indeed, in *World Risk Society* (1999: 23) Beck reverts from objectivism and relativism to the categories of realism and constructivism.

2. For example, it is impossible to tell whether or not a prospective sexual partner is an HIV carrier, or whether a particular piece of beef is contaminated.

3. For example, in Britain, the Royal Society (1992: 94) distinguishes between 'objective' and 'subjective' risks, portraying social risks as identifiable via scientific monitoring and quantification.

4. I do not intend to provide a systematic review of empirical studies into risk perception. For a thorough summary see Krimsky and Golding (1992) or Slovic (2000).

5. This process can be identified in a number of cases, such as the attribution of Aids to the gay community, juvenile delinquency to single mothers and violent crime to black youth.

6. The remaining risks were: food colouring; nuclear power; driving a car; mugging; home accidents; war; ozone depletion and microwave oven usage.

7. Of course, these findings only tell us about the general picture and tend to squash intra and extra-group difference. Naturally, some men will be highly risk-averse and some women will exhibit risk-seeking behaviour.

8. Harking back to Douglas' sentiments, Macgill observed that those who refuted the radioactive risk were often directly related to people dependent upon the plant for their livelihood.

9. See 'Aids – Our Gift to Africa' by Giles Foden in the *Guardian Review*, 30 October 1999: 9.

10. See 'AIDS at Record Levels with 2.6 m Deaths This Year' in the *Daily Telegraph*, 24 November 1999: 3.

11. Statistic cited in 'Social Trends 2002: A Snapshot' in *Society Matters* 5, Open University Bulletin: 3.

12. Beck's position on the distributional logic is perhaps more flexible than Scott lets on. Beck does acknowledge that risk positions can be relative, particularly in the transitional period between industrial and risk society. Take the following quote from *Environmental Politics in an Age of Risk* (Beck, 1995a: 142): 'it may be that everyone is in the same boat in the flood of hazard. But, as is often the case, here there are captains, passengers, helmsmen, engineers and people drowning.'

13. When confronted by certain threats – such as the risk of cancer – an individual may think reflexively and take preventative action. One might, for example, stop smoking and increase consumption of fruit and vegetables. In another situation – such as the risk of sexual disease – the same individual may adopt a more fatalistic approach, rejecting the use of preventative aids.

CHAPTER 6

1. It is important to note that the bulk of Beck's published work on individualisation has been written with his partner Elisabeth Beck-Gernsheim (see Beck and Beck-Gernsheim, 1996; 2002).
2. For example, much has been made of the 'flexibly specialised' workshops in north-eastern Italy in which well trained, multiskilled workers engage in a diverse range of tasks (see Belussi, 1987; Braham, 1997: 157; Murray, 1988).
3. In the British case, the flexibilisation of labour was underpinned by the deregulation of employment law in the 1980s. Under Margaret Thatcher, the Conservative government pushed though a notorious range of anti-union legislation, including the Employment Acts of 1980, 1982, 1988 and 1990. Most overtly, the Trade Union Reform Act of 1993 legalised full postal balloting, a notice period prior to strike action, the removal of ACAS's requirement to encourage collective bargaining and the abolition of the 26 wages councils (see Blyton and Turnbull, 1994: 165).
4. Cited in 'Social Trends 2002: A Snapshot', *Society Matters* 5, Open University Bulletin: 3.
5. In this regard, the growth of fee paying at universities is likely to encourage more students to attend local study centres and to remain in the family home.
6. In recent work, Beck acknowledges that the coverage of individualisation will differ according to cultural conditions (Beck and Beck-Gernsheim, 2002: 5). However, the knock-on effects of this for the risk society framework remain unexplained.

CHAPTER 7

1. The phrase originally belongs to Paul Tillich (1952) and is appositely related to Beck's work by Iain Wilkinson (2001).
2. *Twister* was first screened in Britain by BBC1 on Saturday 3 July 1999.
3. For instance, people suffering from temporary ailments can often be heard bemoaning the fact that they had been 'taking their vitamins' or 'eating well' prior to the onset of illness.
4. Lash claims that his aesthetic version of reflexivity is 'more rooted, more foundational – more situated in a *Sittlichkeit* of social nature' (Lash, 1993: 10).
5. For example, a member of the 'lay public' who works as a plumber may be dependent upon a 'governmental expert' for advice about the risk of deep vein thrombosis. However, if a water pipe bursts in the 'expert's home, the 'layperson' becomes the 'expert' whose knowledge is required to prevent the house from flooding.

CHAPTER 8

1. A more detailed critique of the of the formal political process can be found in Claus Offe's classic, *Contradictions of the Welfare State* (1984) or Conrad Lodziak's *Manipulating Needs: Capitalism and Culture* (1995).

2. The connectivity between the local and the global has been rigorously documented elsewhere. See, for example, John Tomlinson's *Globalization and Culture* (1999), or David Held's *A Globalizing World? Culture, Economics, Politics* (2000).

3. Interestingly, Beck's original title for *The Reinvention of Politics* – first published in Germany in 1992 – was *Beyond Left and Right*. The same title was subsequently adopted by Giddens (1994).

4. Of course, a decline in interest in the formal process must be set within the wider context of the burgeoning global scope of multinational corporations and the diminishing power of nation states as political power blocs (see Held, 2000).

5. On 15 February 2003 the largest political demonstration in British history took place with an estimated 2 million people marching against the war on Iraq.

6. For example, in Britain, the Food Standards Agency was founded in response to public concerns about a lack of clear direction and advice about food safety.

7. It must be noted that many of the risks to public health which have been attributed to genetically modified foods are yet to materialise.

8. A BBC documentary entitled 'Is GM Safe?' suggested that in 1996 British supermarkets were stocking an average of 2,000 genetically modified products, as compared to a current average of less than 100.

9. Given the abstract nature of discourse, this is an extremely difficult proposition to verify at an empirical level.

10. For example, in Britain the problem of monitoring the activity of paedophiles in the community led to the *News of the World* publishing the names and photos of convicted paedophiles. This practice effectively individualised the risk and led to aggressive vigilante campaigns around the country. These campaigns reached their apex when a paediatrician was mistaken for a paedophile and became the subject of violence and abuse.

11. However, both public and governmental attitudes towards blame will be influenced by the content and context of risk. Blame for crime is more easily individualised than responsibility for nuclear or chemical pollution. The extent to which risk conflicts are individualised or collectivised will in turn shape the degree of political opposition or acquiescence. Certain risks may promote institutional critique, others can be utilised to mask institutional culpability.

12. McGrew (1999: 138) notes that the number of intergovernmental national organisations had risen from 37 in 1909 to 300 in 1999.

13. Indeed, a cursory trawl around the world wide web reveals that vitriolic racist groups and Machiavellian business organisations are adept at forming their own subpolitical groups.

CONCLUSION

1. For instance, in 1952 the infamous 'pea-souper' smog which killed 4,500 Londoners bore all the hallmarks of a manufactured risk (Macionis and Plummer, 1999: 650). Half a century later, flood, famine and

earthquakes continue to have ruinous effects across tracts of Africa, Asia and South America.

2. The German term for 'story-teller' is *Schriftsteller*. As Lash and Wynne (1992) note, this term has no real equivalent in the English language, being rather unsatisfactorily translated as 'essayist' or non-fiction writer.

3. In its underlying modernism, the risk society thesis has also attracted comparison with Habermas's investigation into the public sphere (see Culpitt, 1999: 137; Lash and Wynne, 1992: 8; McGuigan, 1999: 130).

4. For example, the Eurocentrism implicit in Beck's risk society thesis has opened up the space for more subtle cross-cultural approaches to blossom (see Bujra, 2000; Skinner, 2000).

Bibliography

Abbinnett, R. (2000) 'Science, Technology and Modernity: Beck and Derrida on the Politics of Risk', *Cultural Values*, 4 (1): 101–26

Adam, B. (1998) *Timescapes of Modernity: The Environment and Environmental Hazards* London: Routledge

—— (2000a) 'The Temporal Gaze: The Challenge for Social Theory in the Context of GM Food', *British Journal of Sociology*, 51 (1): 125–42

—— (2000b) 'The Media Timescapes of BSE News' in S. Allan, B. Adam and C. Carter (eds) *Environmental Risks and the Media* London: Routledge: 117–29

—— and van Loon, J. (2000) 'Repositioning Risk: The Challenge for Social Theory' in B. Adam, U. Beck and J. van Loon (eds) *The Risk Society and Beyond* London: Sage: 1–33

——, Beck, U. and van Loon, J. (2000) *The Risk Society and Beyond* London: Sage

Alexander, J. and Smith, P. (1996) 'Social Science and Salvation: Risk Society as Mythic Discourse', *Zeitschrift für Soziologie*, 25 (4): 251–62

Allan, S., Adam, B. and Carter, C. (2000) 'Introduction: The Media Politics of Environmental Risk' in S. Allan, B. Adam and C. Carter (eds) *Environmental Risks and the Media* London: Routledge: 1–26

Allen, J. (2000) 'Power: Its Institutional Guises' in G. Hughes and R. Fergusson (eds) *Ordering Lives: Family, Work and Welfare* London: Routledge: 7–44

Anderson, A. (1997) *Media, Culture and the Environment* London: UCL Press

—— (2000) 'Environmental Pressure Politics and the Risk Society' in S. Allan, B. Adam and C. Carter (eds) *Environmental Risks and the Media* London: Routledge: 93–104

Balkin, S. (1979) 'Victimization Rates, Safety and Fear of Crime', *Social Problems*, 26 (2): 343–58

Barnes, M. (1999) *Building a Deliberative Democracy: An Evaluation of Two Citizens Juries* London: IPPR

Barwise, P. and Gordon, D. (1998) 'The Economics of the Media' in A. Briggs and P. Cobley (eds) *The Media: An Introduction* London: Longman: 192–209

Bauman, Z. (1991) *Modernity and Ambivalence* Cambridge: Polity Press

—— (1992) *Imitations of Post-Modernity* London: Routledge

Beck, U. (1987) 'The Anthropological Shock: Chernobyl and the Contours of the Risk Society', *Berkley Journal of Sociology*, 32: 153–65

—— (1992) *Risk Society: Towards a New Modernity* London: Sage

—— (1994) 'The Reinvention of Politics: Towards a Theory of Reflexive Modernization' in U. Beck, A. Giddens and S. Lash (eds) *Reflexive Modernization: Politics, Tradition and Aesthetics in the Modern Social Order* Cambridge: Polity Press: 1–55

—— (1995a) *Ecological Politics in an Age of Risk* Cambridge: Polity Press

—— (1995b) 'Freedom for Technology', *Dissent*, Fall Edition: 503–7

—— (1996a) 'World Risk Society as Cosmopolitan Society? Ecological Questions in a Framework of Manufactured Uncertainties', *Theory, Culture and Society*, 13 (4): 1–32

—— (1996b) 'Risk Society and the Provident State' in B. Szerszinski, S. Lash and B. Wynne (eds) *Risk, Environment and Modernity: Towards a New Ecology* London: Sage: 27–43

—— (1997) *The Reinvention of Politics: Rethinking Modernity in the Global Social Order* Cambridge: Polity Press

—— (1998a) *Democracy Without Enemies* Cambridge: Polity Press

—— (1998b) 'Politics of Risk Society' in J. Franklin (ed.) *The Politics of Risk Society* Cambridge: Polity Press: 9–22

—— (1999) *World Risk Society* Cambridge: Polity Press

—— (2000a) *Brave New World of Work* Cambridge: Polity Press

—— (2000b) *What Is Globalization?* Cambridge: Polity Press

—— (2000c) 'The Cosmopolitan Perspective: Sociology of the Second Age of Modernity', *British Journal of Sociology*, 51 (1): 79–105

—— (2000d) 'Risk Society Revisited: Theory, Politics and Research Programmes' in B. Adam, U. Beck and J. van Loon (eds) *The Risk Society and Beyond: Critical Issues for Social Theory* London: Sage: 211–29

—— (2000e) 'Foreword' in S. Allan, B. Adam and C. Carter (eds) *Environmental Risks and the Media* London: Routledge: xi–xiv

—— (2002) 'On World Risk Society', *Logos*, 1 (4): 1–18

—— and Beck-Gernsheim, E. (1995) *The Normal Chaos of Love* London: Polity Press

—— and Beck-Gernsheim, E. (1996) 'Individualization and Precarious Freedoms: Perspectives and Controversies of a Subject-oriented Sociology' in P. Heelas (ed.) *Detraditionalization: Critical Reflections on Authority and Identity* Oxford: Blackwell: 23–48

—— and Beck-Gernsheim, E. (2002) *Individualization: Institutionalized Individualism and its Social and Political Consequences* London: Sage

—— Giddens, A. and Lash, S. (1994) *Reflexive Modernization: Politics, Tradition and Aesthetics in the Modern Social Order* Cambridge: Polity Press

Beck-Gernsheim, E. (2000) 'Health and Responsibility: From Social Change to Technological Change and Vice Versa' in B. Adam, U. Beck and J. van Loon (eds) *The Risk Society and Beyond* London: Sage: 122–35

Bell, A. (1994) 'Climate of Opinion: Public and Media Discourse on the Global Environment', *Discourse and Society* 5: 33–64

Belussi, F. (1987) 'Benetton Italy: Beyond Fordism and Flexible Specialisation' in S. Mitter (ed.) *Computer Aided Manufacturing: The Clothing Industry in Four European Countries* Berlin: Springer Verlag: 73–91

Bennett, P. (1998) *Communicating about Risks to Public Health: Pointers to Good Practice* London: HMSO

Bernstein, P. (1996) *Against the Gods: The Remarkable Story of Risk* New York: John Wiley

Blake, J. (1999) 'Overcoming the Value-Action Gap in Environmental Policy: Tensions Between National Policy and Local Experience', *Local Environment*, 4 (3): 257–78

Blyton, P. and Turnbull, P. (1994) *The Dynamics of Employee Relations* Basingstoke: Macmillan

Boden, D. (2000) 'Worlds in Action: Information, Instantaneity and Global Futures Trading' in B. Adam, U. Beck and J. van Loon (eds) *The Risk Society and Beyond: Critical Issues for Social Theory* London: Sage: 183–97

Boyne, R. (2003) *Risk* Buckingham: Open University Press

Bradbury, J. (1989) 'The Policy Implications of Differing Concepts of Risk', *Science, Technology and Human Values* 14 (4): 380–99

Braham, P. (1997) 'Fashion: Unpacking a Cultural Production' in P. du Gay (ed.) *Production of Culture/Cultures of Production* London: Sage: 119–76

Bromley, S. (2000) 'Political Ideologies and the Environment' in D. Goldblatt (ed.) *Knowledge and the Social Sciences: Theory, Method, Practice* London: Routledge: 77–118

Bronner, S.E. (1995) 'Ecology, Politics and Risk: The Social Theory of Ulrich Beck', *Capitalism, Nature, Socialism: A Journal of Socialist Ecology*, 6 (1): 67–86

Brynner, J. and Ashford, S. (1994) 'Politics and Participation: Some Antecedents of Young People's Attitudes to the Political System and Political Activity', *European Journal of Social Psychology*, 24: 223–36

Bujra, J. (2000) 'Risk and Trust: Unsafe Sex, Gender and AIDS in Tanzania' in P. Caplan (ed.) *Risk Revisited* London: Pluto Press: 59–84

Burger, J., Pflugh, K., Lurig, L., Von Hagen, L. and Von Hagen, S. (1999) 'Fishing in Urban New Jersey: Ethnicity Affects Information Sources, Perception and Compliance', *Risk Analysis*, 19 (2): 217–29.

Caplan, P. (2000a) *Risk Revisited* London: Pluto Press

—— (2000b) 'Eating British Beef with Confidence: A Consideration of Consumers' Responses to BSE in Britain' in P. Caplan (ed.) *Risk Revisited* London: Pluto Press: 184–203

Castel, R. (1991) 'From Dangerousness to Risk' in G. Burchell, C. Gordon and P. Miller (eds) *The Foucault Effect: Studies in Governmentality* London: Harvester Wheatsheaf: 281–98

Caygill, H. (2000) 'Liturgies of Fear: Biotechnology and Culture' in B. Adam, U. Beck and J. van Loon (eds) *The Risk Society and Beyond* London: Sage: 155–64

Clark, N. (1997) 'Panic Ecology: Nature in the Age of Superconductivity', *Theory, Culture and Society*, 14 (1): 77–96

Cohen, M. (2000) 'Science and Society in Historical Perspective: Implications for Social Theories of Risk', *Environmental Politics*, 8 (2): 153–77

Cohen, S. (1972) *Folk Devils and Moral Panics* London: Paladin

Cohn, S. (2000) 'Risk, Ambiguity and the Loss of Control: How People With a Chronic Illness Experience Complex Biomedical Causal Models' in P. Caplan (ed.) *Risk Revisited* London: Pluto Press: 204–25

Coleman, C. (1995) 'Science, Technology and Risk Coverage of a Community Conflict', *Media, Culture and Society*, 17: 65–79

Collins, R. K. (1992) *Dictating Content: How Advertising Pressure Can Corrupt a Free Press* Washington, DC: Center for the Study of Commercialism

Connell, R.W. (1995) *Masculinities* Cambridge: Polity Press

Coote, A. (1998) 'Risk and Public Policy: Towards a High Trust Democracy' in J. Franklin (ed.) *The Politics of Risk Society* Cambridge: Polity Press: 124–32

—— and Lenaghen, J. (1997) *Citizens' Juries: Theory into Practice* London: IPPR

—— and Mattinson, D. (1997) *Twelve Good Neighbours* London: Fabian Society

Corner, J., Richardson, K. and Fenton, N. (1990a) *Nuclear Reactions: Form and Response in Public Issue Television* London: John Libbey

Corner, J., Richardson, K. and Fenton, N. (1990b) 'Textualizing Risk: TV Discourse and the Issue of Nuclear Energy', *Media, Culture and Society*, 12: 105–24

Cottle, S. (1998) 'Ulrich Beck, Risk Society and the Media', *European Journal of Communications*, 13 (1): 5–32

—— (2000) 'TV News, Lay Voices and the Visualisation of Environmental Risks' in S. Allan, B. Adam and C. Carter (eds) *Environmental Risks and the Media* London: Routledge: 29–44

Courtney, G. and McAleese, I. (1993) *England and Wales Youth Cohort Study* Report on Cohort 5, Sweep 1 Sheffield: Department of Employment

Cragg Ross Dawson Report (2000) *Qualitative Research to Explore Public Attitudes to Food Safety* London: Cragg Ross Dawson

Crook, S. (1999) 'Ordering Risks' in D. Lupton (ed.) *Risk and Sociocultural Theory: New Directions and Perspectives* Cambridge: Cambridge University Press: 160–85

Croteau, D. and Hoynes. W. (2000) *Media / Society: Industries, Images and Audiences* Thousand Oaks: Pine Forge Press

CSEC Report (1997) *Uncertain World: Genetically Modified Organisms, Food and Public Attitudes in Britain* Lancaster: Centre for the Study of Environmental Change

Culpitt, I. (1999) *Social Policy and Risk* London: Sage

Curran, J. and Seaton, J. (1989) *Power Without Responsibility: The Press and Broadcasting in Britain* London: Routledge

Dawson, G. (2000) 'Work: From Certainty to Flexibility' in G. Hughes and R. Fergusson (eds) *Ordering Lives: Family, Work and Welfare* London: Routledge: 81–114

Day, S. (2000) 'The Politics of Risk among London Prostitutes' in P. Caplan (ed.) *Risk Revisited* London: Pluto: 29–58

Dean, M. (1999) 'Risk, Calculable and Incalculable' in D. Lupton (ed.) *Risk and Sociocultural Theory: New Directions and Perspectives* Cambridge: Cambridge University Press: 131–59

Denscombe, M. (1988) *Sociology Update* Leicester: Olympus

Dickens, P. (1996) *Reconstructing Nature: Alienation, Emancipation and the Division of Labour* London: Routledge

Dion, K., Baren, R. and Miller, N. (1971) 'Why do Groups Make Riskier Decisions than Individuals', *Experimental Social Psychology*, 5: 12–22

Doogan, K. (2001) 'Insecurity and Long-Term Employment', *Work, Employment and Society*, 15 (3): 419–42

Douglas, M. (1966) *Purity and Danger: An Analysis of the Concepts of Pollution and Taboo* London: Routledge

—— (1982) *Natural Symbols: Explorations in Cosmology* New York: Pantheon

—— (1985) *Risk Acceptability According to the Social Sciences* New York: Russell Sage

—— (1992) *Risk and Blame: Essays in Cultural Theory* London: Routledge

—— and Wildavsky, A. (1982) *Risk and Culture: An Essay on the Selection of Technological and Environmental Dangers* Berkeley: University of California Press

Dunwoody, S and Peters, H.P. (1993) 'The Mass Media and Risk Perception' in Ruck, B. (ed.) *Risk is a Construct* Munich: Knesebeck: 14–28

Draper, E. (1993) 'Risk Society and Social Theory', *Contemporary Sociology: A Journal of Reviews*, 22: 641–4

Dryzek, J. (1995) 'Toward an Ecological Modernity', *Contemporary Sociology: A Journal of Reviews*, 28: 231–42

du Gay, P., Hall, S., Janes, L., Mackay, H. and Negus, K. (1997) *Doing Cultural Studies: The Story of the Walkman* London: Sage

Eagleton, T. (2003) *The Gatekeeper* London: Penguin

Elliott, A. (2002) 'Beck's Sociology of Risk: A Critical Assessment', *Sociology*, 36 (2): 293–315

Elridge, J. (1999) 'Risk, Society and the Media: Now You See it, Now You Don't' in G. Philo (ed.) *Message Received: Glasgow Media Group Research 1993–1998* New York: Longman: 106–27

Emslie, C., Hunt, K. and Macintyre, S. (2000) 'Gender or Job Differences? Working Conditions amongst Men and Women in White Collar Occupations', *Work, Employment and Society*, 13 (4): 711–29

Engel, U and Strasser, H. (1998) 'Notes on the Discipline', *Canadian Journal of Sociology*, 23 (1): 91–103

ESRC Global Environmental Change Report (1999) *The Politics of GM Foods: Risk, Science and Public Trust* Special Briefing No 5 Brighton: University of Sussex

Ewald, F. (1986) *L'Etat Providence* Paris: Grasser and Fasquell

—— (1991) 'Insurance and Risk' in G. Burchell, C. Gordon and P. Miller (eds) *The Foucault Effect: Studies in Governmentality* London: Harvester Wheatsheaf: 197–210

—— (1993) 'Two Infinities of Risk' in B. Massumi (ed.) *The Politics of Everyday Fear* Minneapolis: University of Minnesota Press

Field, J. and Schreer, G. (2000) 'Age Differences in Personal Risk Perceptions', *Risk: Health, Safety and Environment*, 11 (4): 287–95

Finucane, M.L., Slovic, P., Mertz, C., Flynn, J. and Satterfield, T. (2000) 'Gender, Race and Perceived Risk: The White Male Effect', *Health, Risk and Society*, 2 (2): 159–72

Fischhoff, B. (1995) 'Risk Perception and Communication Unplugged: Twenty Years of Process', *Risk Analysis*, 15 (2): 137–45

Flynn, J., Slovic, P. and Mertz, C. (1994) 'Gender, Race and Perception of Environmental Health Risks', *Risk Analysis*, 14 (6): 101–8

Follain, J. (2001) 'Italian Doctor Prepares to Clone First Human Being', *Sunday Times*, 28 January: 30

Foucault, M. (1978) *The History of Sexuality* Harmonsworth: Penguin

—— (1980) *Power/Knowledge* Brighton: Harvester

—— (1991) 'Governmentality' in G. Burchell, C. Gordon and P. Miller (eds) *The Foucault Effect: Studies in Governmentality* London: Harvester Wheatsheaf: 87–104

Fox, N. (1999) 'Postmodern Reflections on "Risk", "Hazards" and "Life Choices"' in D. Lupton (ed.) *Risk and Sociocultural Theory: New Directions and Perspectives* Cambridge: Cambridge University Press: 12–33

Franklin, J. (1998) *The Politics of Risk Society* Cambridge: Polity Press

Friedman, S., Villamil, K., Suriano, R and Egolf, B. (1996) 'Alar and Apples: Newspapers, Risk and Media Responsibility', *Public Understandings of Science*, 5: 1–20

Furedi, F. (1997) *Culture of Fear: Risk Taking and the Morality of Low Expectation* London: Cassell

Furlong, A. and Cartmel, F. (1997) *Young People and Social Change: Individualization and Risk in Late Modernity* Buckingham: Open University Press

—— and Cartmel, F. (1998) 'Growing Up in the Risk Society', *Sociology Review*, November Edition: 27–30

Galtung, J. and Ruge, M. (1974) 'The Structure of Foreign News' in J. Tunstall (ed.) *Media Sociology* London: Constable: 259–300

Gangl, M. (2001) 'Changing Labour Markets and Early Career Outcomes: Labour Market Entry in Europe Over the Past Decade', *Work, Employment and Society*, 16 (1): 67–90

Gardner, C.B. (1995) *Passing By: Gender and Public Harassment* Berkeley: University of California Press

Giddens, A. (1991) *Modernity and Self Identity* Cambridge: Polity Press

—— (1994) *Beyond Left and Right: The Future of Radical Politics* Cambridge: Polity Press

—— (1998) 'Risk Society: The Context of British Politics' in J. Franklin (ed.) *The Politics of Risk Society* Cambridge: Polity Press: 23–34

—— (1999) *The Reith Lectures: Risk* BBC News Online <http://news.bbc.co.uk. reith_99>

Glasner, P. (2000) 'Reporting Risks, Problematising Public Participation and the Human Genome Project' in S. Allan, B. Adam and C. Carter (eds) *Environmental Risks and the Media* London: Routledge: 130–41

Goldblatt, D. (1995) *Social Theory and the Environment* Cambridge: Polity Press

Goldthorpe, J. and Marshall, G. (1992) 'The Promising Future of Class Analysis', *Sociology*, Annual Collection: 381–400

Gorz, A. (1982) *Farewell to the Working Class: An Essay on Post-Industrial Socialism* London: Pluto Press

Gorz, A. (1988) *Critique of Economic Reason* London: Verso

—— (1994) *Capitalism, Socialism, Ecology* London: Verso

—— (2000) *Reclaiming Work: Beyond the Wage Based Society* Oxford: Basil Blackwell

Graham, J. and Clemente, K. (1996) 'Hazards in the News: Who Believes What?', *Risk in Perspective*, 4 (4): 1–4

Green, P. (2003) 'American Television, Crime and the Risk Society' in K. Stenson and R. Sullivan (eds) *Crime, Risk and Justice* Cullompton: Willan Publishing: 214–27

Green, T. (2003) 'The Shock of the True', *Independent Magazine*, 14 June: 8

Greenberg, M., Sachsburg, D., Sandman, P. and Salome, K. (1989) 'Network Evening News Coverage of Risk', *Risk Analysis*, 9: 119–26

Grice, A. (2003) 'Labour Has Failed to End Class Divisions, Ministers Admit', *Independent*, 4 January: 6

Grove-White, R. (1998) 'Risk Society, Politics and BSE' in J. Franklin (ed.) *The Politics of Risk Society* Cambridge: Polity Press: 50–3

Gustafson, P.E. (1998) 'Gender Differences in Risk Perception: Theoretical and Methodological Perspectives', *Risk Analysis*, 18 (6): 805–11

Habermas, J. (1989) *The Theory of Communicative Action* Cambridge: Polity Press

Hajer, M. and Kesselring, S. (1999) 'Democracy in the Risk Society? Learning from the New Politics of Mobility in Munich', *Environmental Politics*, 8 (3): 1–23

Hall, S. (1973) 'A World at One With Itself' in S. Cohen and J. Young (eds) *The Manufacture of News* London: Constable: 147–56

—— (1980) 'Encoding/Decoding' in S. Hall., D. Hobson., A. Lowe and P. Willis (eds) *Culture, Media, Language* Birmingham: Hutchison: 128–38

——, Crichter, C., Jefferson, T. and Roberts, B. (1982) *Policing the Crisis* London: Macmillan

Handmer, J. (1995) 'Communicating Uncertainty: Perspectives and Themes' in T. Norton., T. Beer and S. Dovers (eds) *Risk Uncertainty in Environmental Management* Canberra: Australian Academy of Science: 101–17

Hansen, A. (1990) *The News Construction of the Environment: A Comparison of British and Danish Television News* Leicester: Centre for Mass Communications Research

—— (1991) 'The Media and the Social Construction of the Environment', *Media, Culture and Society*, 13: 443–58

—— (2000) 'Claims-Making and Framing in British Newspaper Coverage of the 'Brent Spar' Controversy' in S. Allan, B. Adam and C. Carter (eds) *Environmental Risks and the Media* London: Routledge: 55–72

Hargreaves, I. (2000) *Who's Misunderstanding Whom? An Inquiry into the Relationship Between Science and the Media* London: ESRC

Harris, P. and O'Shaughnessy, N. (1997) 'BSE and Marketing Communication Myopia: Daisy and the Death of the Sacred Cow', *Risk Decision and Policy*, 2 (1): 29–39

Heelas, P. (ed.) (1996) *Detraditionalization: Critical Reflections on Authority and Identity* Oxford: Blackwell

Held, D. (2000) *A Globalizing World? Culture, Economics, Politics* London: Routledge

——, McGrew, A., Goldblatt, D. and Perraton, J. (1999) *Global Transformations: Politics, Economics and Culture* Cambridge: Polity Press

Hillyard, P. and Percy-Smith, C. (1988) *The Coercive State* London: Macmillan

Hinchcliffe, S. (2000) 'Living with Risk: The Unnatural Geography of Environmental Crises' in S. Hinchcliffe and K. Woodward (eds) *The Natural and the Social: Uncertainty, Risk, Change* London: Routledge: 117–54

Ho, M.W. (1997) 'The Unholy Alliance', *The Ecologist*, 27 (4): 152–8

Hughes, G. and Fergusson, R. (2000) *Ordering Lives: Family, Work and Welfare* London: Routledge

Hutton, W. (1996) *The State We're In* London: Vintage

Irwin, A. (1989) 'Deciding about Risk: Expert Testimony and the Regulation of Hazard' in J. Brown (ed.) *Environmental Threats: Analysis, Perception, Management* London: Belhaven: 19–32

——, Allan, S. and Welsh, I. (2000) 'Nuclear Risks: Three Problematics' in B. Adam, U. Beck and J. van Loon (eds) *The Risk Society and Beyond: Critical Issues for Social Theory* London: Sage: 78–104

IPCC Report (2001) *Climate Change 2001: The Scientific Basis* Cambridge: Cambridge University Press

Jackson, S. and Scott, S. (1999) 'Risk Anxiety and the Social Construction of Childhood' in D. Lupton (ed.) *Risk and Sociocultural Theory: New Directions and Perspectives* Cambridge: Cambridge University Press: 86–107

Jasanoff, S. (1999) 'The Songlines of Risk', *Environmental Politics*, 9 (2): 135–53

Joffe, H. (1999) *Risk and the Other* Cambridge: Cambridge University Press

Jones, A., Pettiford, L., Smith, R. and Tomlinson, J. (1997) *Euro-Trash: Television and Environmental Campaigns in Three European Countries* CRICC Report, Nottingham: Nottingham Trent University

Jones, G. (1995) *Leaving Home* Buckingham: Open University Press

Kamppinen, M. and Wilenius, M. (2001) 'Risk Landscapes in the Era of Social Transition', *Futures*, 33: 307–17

Kasperson, R. and Kasperson, J. (1996) 'The Social Amplification and Attenuation of Risk', *Annals of the American Academy of Political and Social Science*, 545: 116–25

Kerr, A. and Cunningham-Burley, S. (2000) 'On Ambivalence and Risk: Reflexive Modernity and the New Human Genetics', *Sociology*, 34 (2): 283–304

Knight, F. (1921) *Risk, Uncertainty and Profit* Boston, MA: Houghton Mifflin

Krimsky, S. and Golding, D. (1992) *Social Theories of Risk* London: Praeger

Langford, I., Marris, C., and O'Riordan, T. (1999) 'Public Reactions to Risk: Social Structures, Images of Science and the Role of Trust' in P. Bennett and K. Calman (eds) *Risk Communication and Public Health* Oxford: Oxford University Press

Lash, S. (1990) *Sociology of Post-modernism* London: Routledge

—— (1993) 'Reflexive Modernization: The Aesthetic Dimension', *Theory, Culture and Society*, 10: 1–23

—— (1994) 'Reflexivity and its Doubles: Structure, Aesthetics, Community' in U. Beck, A. Giddens and S. Lash (eds) *Reflexive Modernization: Politics, Tradition and Aesthetics in the Modern Social Order* Cambridge: Polity Press: 110–73

—— (2000) 'Risk Culture' in B. Adam, U. Beck and J. van Loon (eds) *The Risk Society and Beyond: Critical Issues for Social Theory* London: Sage: 45–62

—— (2002) 'Individualization in a Non-Linear Mode' in U. Beck and E. Beck-Gernsheim *Individualization: Institutionalized Individualism and its Social and Political Consequences* London: Sage: vi–xiii

—— and Urry, J. (1994) *Economies of Signs and Space* London: Sage

—— and Wynne, B. (1992) 'Introduction' in U. Beck, *Risk Society: Towards a New Modernity* London: Sage

——, Szerszinski, B. and Wynne, B. (1996) *Risk, Environment and Modernity: Towards a New Ecology* London: Sage

Laurance, J. (2000) 'Highly Infectious: Scare Stories', *Independent*, 2 November: 5

Lazo, J., Kinnell, J. and Fisher, A. (2000) 'Expert and Layperson Perceptions of Ecosystem Risk', *Risk Analysis*, 20 (4): 179–203

Lee, T.R. (1981) 'Perceptions of Risk: The Public's Perception of Risk and the Question of Irrationality', *Proceedings of The Royal Society* London: Royal Society: 5–16

Leiss, W. (2000) 'Book Review: Risk Society, Towards a New Modernity', *Canadian Journal of Sociology Online*, 25 (3) <www.ualberta.ca/~cjscopy/articles/leiss.html>

Lichfield, J. (2003) 'Protestors to be Kept 30 Miles Away', *Independent*, 31 May 2003: 2

Linne, O. and Hansen, A. (1990) *News Coverage of the Environment: A Comparative Study of Journalistic Practices and Television Presentation in Denmark's Radio and the BBC* Research Report No.1B/90 Copenhagen: Danmarks Radio

Lippmann, W. (1965) *Public Opinion* New York: Free Press

Lodziak, C. (1986) *The Power of Television: A Critical Appraisal* London: Pinter

—— (1995) *Manipulating Needs: Capitalism and Culture* London: Pluto Press

—— (2002) *The Myth of Consumerism* London: Pluto Press

Luhmann, N. (1993) *Risk: A Sociological Theory* New York: Aldine/de Gruyter

Lupton, D. (1995) *The Imperative of Health: Public Health and the Regulated Body* London: Sage

—— (1999a) *Risk* London: Routledge

—— (ed.) (1999b) *Risk and Sociocultural Theory: New Directions and Perspectives* Cambridge: Cambridge University Press

—— and Tulloch, J. (2002a) 'Life Would Be Pretty Dull Without Risk: Voluntary Risk Taking and Its Pleasures', *Health, Risk and Society*, 4 (2): 113–24

—— and Tulloch, J. (2000b) 'Risk is Part of Your Life: Risk Epistemologies Amongst a Group of Australians', *Sociology*, 36 (2): 317–34

Lyng, S. (1990) 'Edgework: A Social Psychological Analysis of Voluntary Risk Taking', *American Journal of Sociology*, 95 (4): 851–6

Macgill, S. (1989) 'Risk Perception and the Public: Insights from Research Around Sellafield' in J. Brown (ed.) *Environmental Threats: Perception, Analysis and Management* London: Belhaven Press: 48–66

Macintyre, S., Reilly, J., Miller, D. and Eldridge, J. (1998) 'Food Choice, Food Scares and Health: The Role of the Media' in A. Murcott (ed.) *The Nation's Health: The Social Science of Food Choice* London: Longman

Macionis, J. and Plummer, K. (1998) *Sociology: A Global Introduction* Upper Saddle River: Prentice-Hall

Mackey, E. (1999) 'Constructing an Endangered Nation: Risk, Race and Rationality in Australia's Native Title Debate' in D. Lupton (ed.) *Risk and Sociocultural Theory: New Directions and Perspectives* Cambridge: Cambridge University Press: 108–30

Mackintosh, M. and Mooney, G. (2000) 'Identity, Inequality and Social Class' in K. Woodward (ed.) *Questioning Identity: Gender, Class, Nation* London: Routledge: 79–114

Macnaghtan, P. (2003) 'Embodying the Environment in Everyday Life Practices', *The Sociological Review*, 51 (1): 63–84

—— and Urry, J. (1998) *Contested Natures* London: Sage

Mann, M. (1982) 'The Social Cohesion of Liberal Democracy' in A. Giddens and D. Held (eds) *Classes, Power and Conflict: Classical and Contemporary Debates* London: Macmillan: 331–94

Marcuse, H. (1964) *One Dimensional Man* London: Sphere

Marris, C. and Langford, I. (1996) 'No Cause for Alarm', *New Scientist*, 28 September: 36–9

Marshall, B.K. (1999) 'Globalisation, Environmental Degradation and Ulrich Beck's Risk Society', *Environmental Values*, 8: 253–75

Marshall, G. (1988) *Social Class in Modern Britain* London: Unwin Hyman

Marwick, A. (1990) *British Society Since 1945* London: Penguin

Matthiessen, P. (1999) 'Get Down to Earth', *Guardian*, 30 October: 14

McCarthy, M. (2003a) 'The Great Rainforest Tragedy', *Independent*, 28 June: 1

—— (2003b) 'Reaping the Whirlwind', *Independent*, 3 July: 1

McGlone, F., Park, A. and Roberts, C. (1996) *Relative Values: Kinship and Friendship* London: HMSO

McGrew, A. (2000) 'Power Shift: From National Governments to Global Governance?' in D. Held (ed) *A Globalizing World: Culture, Economics, Politics* London: Sage: 127–67

McGuigan, J. (1999) *Modernity and Postmodern Culture* London: Open University Press

McMylor, P. (1996) 'Goods and Bads', *Radical Philosophy*, 77: 52–3

McNair, B. (1998) 'Technology: New Technologies and the Media' in A. Briggs and P. Cobley (eds) *The Media: An Introduction* London: Longman: 173–85

Miller, M. and Riechert, B. (2000) 'Interest Group Strategies and Journalistic Norms: News Media Framing of Environmental Issues' in S. Allan, B. Adam and C. Carter (eds) *Environmental Risks and the Media* London: Routledge: 45–55

Mintel Report (1999) *Consumer Attitudes to GM Foods* London: Mintel International Group Limited

Mol, P.J. and Spaargaren, G. (1993) 'Environment, Modernity and the Risk Society: The Apocalyptic Horizon of Environmental Reform', *International Sociology*, 8 (4): 431–59

Mooney, G., Kelly, B., Goldblatt, D. and Hughes, G. (2000) *Tales of Fear and Fascination: The Crime Problem in the Contemporary UK* London: Routledge

Morley, D. (1980) *The Nationwide Audience: Structure and Encoding* London: BFI

Munton, R. (1997) 'Engaging Sustainable Development: Some Observations on Progress in the UK', *Progress in Human Geography*, 21: 147–63

Murray, R. (1988) 'Life after Ford', *Marxism Today* October Edition: 8–13

Mythen, G. (2002) 'Communicating Risk: Reconfiguring Expert–Lay Relations' in F. Redmill and T. Anderson (eds) *Components of System Safety* London: Springer: 196–213

——, Wales, C., French, S. and Maule, J. (2000) *Risk Communication and Risk Perception: A Critical Review* MBS Working Paper, 411: Manchester: University of Manchester

Negrine, R. (1994) *Politics and the Mass Media in Britain* London: Routledge

Negus, K. (1997) 'The Production of Culture' in P. du Gay (ed.) *The Production of Culture/Cultures of Production* London: Sage: 67–118

Nelkin, D. (1987) *Selling Science: How the Press Covers Science and Technology* New York: Freeman

North, R.D. (1997) 'Life's a Gamble, It's True', *Independent*, 15 November: 21

Nugent, S. (2000) 'Good Risk, Bad Risk: Reflexive Modernisation and Amazonia' in P. Caplan (ed.) *Risk Revisited* London: Pluto Press: 226–47

O'Malley, P. (2001) 'Policing Crime Risks in the Neo-Liberal Era', in K. Stenson and R. Sullivan (eds) *Crime, Risk and Justice* Cullompton: Willan Publishing: 89–103

Offe, C. (1984) *Contradictions of the Welfare State* London: Macmillan

—— (1994) *Modernity and the State: East, West* Cambridge: Polity Press

ONS (2001) *Social Trends 31* London: The Stationery Office

Palmer, J. (1998) 'News Values' in A. Briggs and P. Cobley (eds) *The Media: An Introduction* London: Longman: 377–92

Park, A. (1996) 'Teenagers and their Politics' in R. Jowell, J. Curtice, A. Park, L. Brook and D. Ahrendt (eds) *British Social Attitudes Survey* 12th Report Aldershot: Dartmouth

Perrons, D. (2000) 'Living with Risk: Labour Market Transformation, Employment Policies and Social Reproduction in the UK', *Economic and Industrial Democracy*, 21: 283–310

Phillips Report: The BSE Inquiry (2000) London: Stationery Office

Philmore, P. and Moffatt, S. (2000) 'Industry Causes Lung Cancer: Would You Be Happy With That Headline?' in S. Allan, B. Adam and C. Carter (eds) *Environmental Risks and the Media* London: Routledge: 105–16

Philo, G. (1999) *Message Recieved: Glasgow Media Group Research 1993–1998* Cambridge: Cambridge University Press

Pidgeon, N. (2000) 'Take a Chance', *New Scientist*, 12 August: 46–7

Polanyi, K. (1975) *The Great Transformation* New York: Octagon Books

Prior, L., Glasner, P. and McNally, R. (2000) 'Genotechnology: Three Challenges to Risk Legitimation' in B. Adam, U. Beck and J. van Loon (eds) *The Risk Society and Beyond: Critical Issues for Social Theory* London: Sage: 105–21

Purdue, D. (1995) 'Whose Knowledge Counts? Experts, Counter-Experts and the Lay Public', *The Ecologist*, 25: 170–2

Ratzan, S. (1998) *The Mad Cow Crisis: Health and the Public Good* London: UCL Press

Ravetz, J., Funtowicz, S. and Irwin, A. (1989) 'Conclusion and Policy Perspectives' in J. Brown (ed.) *Environmental Threats: Perception, Analysis and Management* London: Belhaven: 133–7

Reilly, J. (1998) 'Food Choice, Food Scares and Health: The Role of the Media', in A. Murcott (ed.) *The Nations Diet: The Social Science of Food Choice* London: Longman: 44–59

—— (1999) 'Just Another Food Scare? Public Understanding of the BSE Crisis' in G. Philo (ed.) *Message Received: Glasgow Media Group Research 1993–1998* New York: Longman: 128–46

—— and Kitzinger, J. (1997) 'The Rise and Fall of Risk Reporting: Media Coverage of Human Genetics Research, False Memory Syndrome and Mad Cow Disease', *European Journal of* Communication, 12 (3): 319–50

Reiner, R., Livingstone, S. and Allen, J. (2003) 'Casino Culture: Media and Crime in a Winner-Loser Society' in K. Stenson and R. Sullivan (eds) *Crime, Risk and Justice* Cullompton: Willan Publishing: 175–93

Robertson, R. (1992) *Globalization: Social Theory and Global Culture* London: Sage

Rose, H. (2000) 'Risk, Trust and Scepticism in the Age of the New Genetics' in B. Adam, U. Beck and J. van Loon (eds) *Risk Society and Beyond: Critical Issues for Social Theory* London: Sage: 63–77

Royal Society Report (1992) *Risk, Analysis, Perception and Management* London: Amber

Rustin, M. (1994) 'Incomplete Modernity: Ulrich Beck's Risk Society', *Dissent*, Fall: 394–400

Schilling, C. (1997) 'The Body and Difference' in K. Woodward (ed.) *Identity and Difference* London: Sage: 63–120

Schlesinger, P. (1990) 'Rethinking the Sociology of Journalism: Source Strategies and the Limits of Media Centrism' in M. Ferguson (ed.) *Public Communication: The New Imperatives* London: Sage: 61–83

Scott, A. (2000) 'Risk Society or Angst Society? Two Views of Risk, Consciousness and Community' in B. Adam, U. Beck and J. van Loon (eds) *The Risk Society and Beyond: Critical Issues for Social Theory* London: Sage: 33–46

Scott, S., Jackson, S. and Backett-Milburn, K. (1998) 'Swings and Roundabouts: Risk Anxiety and the Everyday Worlds of Children', *Sociology*, 32: 689–705

Segal, L. (1997) 'Sexualities: The Body and Difference' in K. Woodward (ed.) *Identity and Difference* London: Sage: 183–238

Sherratt, N. and Hughes, G. (2000) 'Family: From Tradition to Diversity?' in G. Hughes and R. Fergusson (eds) *Ordering Lives: Family, Work and Welfare* London: Routledge 45–80

Singer, E. and Endreny, P. (1987) 'Reporting Hazards: Their Benefits and Costs', *The Journal of Communication*, 37: 10–16

Sjöberg, L. and Wahlberg, A. (1997) 'Risk Perception and the Media: A Review of Research on Media Influence on Public Risk Perception', *Rhizikon: Risk Research Reports* Stockholm: Centre for Risk Research

Skeat, W. (1910) *An Etymological Dictionary of the English Language* Oxford: Clarendon

Skinner, J. (2000) 'The Eruption of Chances Peak, Montserrat and the Narrative Containment of Risk' in P. Caplan (ed.) *Risk Revisited* London: Pluto Press: 156–83

Slovic, P. (1987) 'Perception of Risk', *Science*, 236: 280–5

—— (1992) 'Perception of Risk: Reflections on the Psychometric Paradigm' in S. Krimsky and D. Golding (eds) *Social Theories of Risk* Westport: Praeger: 117–52

—— (1993) 'Perceived Risk, Trust and Democracy', *Risk Analysis*, 13 (6): 675–82

—— (2000) *The Perception of Risk* London: Earthscan

——, Lichtenstein, S. and Fischhoff, B. (1981) 'Perceived Risk: Psychological Factors and Social Implications', *Proceedings of the Royal Society* London: Royal Society: 17–34

Smith, B. and Goldblatt, D. (2000) 'Whose Health Is It Anyway?' in S. Hinchcliffe and K. Woodward (eds) *The Natural and the Social: Uncertainty, Risk, Change* London: Sage: 43–77

Smith, G. and Wales, C. (2000) 'Citizens' Juries and Deliberative Democracy', *Political Studies*, 48 (1): 51–65.

Smith, K. (2001) 'The Risk Transition in Developing Countries' in J. Kasperson and R. Kasperson (eds) *Global Environmental Risk* London: Earthscan: 148–73

Smith, M., Law, A., Work, H. and Panay, A. (1997) 'The Reinvention of Politics: Ulrich Beck and Reflexive Modernity', *Environmental Politics* 8 (3): 169–73

Social Trends (1999) London: HMSO

Social Trends (2001) London: HMSO

Social Trends (2002) London: HMSO

Sparks, R. (2003) 'Bringin' it All Back Home: Populism, Media Coverage and the Dynamics of Locality and Globality in the Politics of Crime Control' in K. Stenson and R. Sullivan (eds) *Crime, Risk and Justice* Cullompton: Willan Publishing: 194–213

Stallings, R.A. (1990) 'Media Discourse and the Social Construction of Risk', *Social Problems*, 31: 23–31

Stanko, E. (1996) 'Warnings to Women: Police Advice and Women's Safety in Britain', *Violence Against Women*, 2: 5–24

Stevenson, N. (1999) *The Transformation of the Media: Globalisation, Morality and Ethics* London: Longman

Strydom, P. (1999) 'Hermenetic Culturalism and Its Double: A Key Problem in the Reflexive Modernization Debate', *European Journal of Social Theory*, 2 (1): 45–69

—— (2003) *Risk, Environment and Society* Buckingham: Open University Press

Szerszynski, B., Lash, S. and Wynne, B. (1996) *Risk, Environment and Modernity: Towards a New Ecology* London: Sage

Szerszynski, B. and Toogood, M. (2000) 'Global Citizenship, the Environment and the Media' in S. Allan, B. Adam and C. Carter (eds) *Environmental Risks and the Media* London: Routledge: 218–28

Tacke, V. (2001) 'BSE as an Organisational Construction: A Case Study on the Globalization of Risk', *British Journal of Sociology*, 52 (2): 293–312

Taig, T. (1999) 'Benchmarking in Government: Case Studies and Principles' in P. Bennett and K. Calman, (eds) *Risk Communication and Public Health* Oxford: Oxford University Press: 117–32

Tansey, J. and O'Riordan, T. (1999) 'Cultural Theory and Risk: A Review', *Health, Risk and Society*, 1: 71–90

Taylor, S.E. and Brown, J.D. (1988) 'Illusion and Well Being: A Social Psychological Perspective on Mental Health', *Psychological Bulletin*, 103: 193–210

Taylor-Gooby, P. (1999) 'Risk and the Welfare State', *British Journal of Sociology*, 50 (2): 177–94

Thompson, K. (1997) 'Regulation, De-Regulation and Re-Regulation' in K. Thompson (ed.) *The Media and Cultural Regulation* London: Sage: 9–52

Thompson, M. (1989) 'Engineering and Anthropology: Is There a Difference?' in J. Brown (ed.) *Environmental Threats: Perception, Analysis and Management* London: Belhaven: 138–50

Tillich, P. (1952) *The Courage to Be* Glasgow: Collins

Tomlinson, J. (1997) 'Internationalism, Globalization and Cultural Imperialism' in K. Thompson (ed.) *Media and Cultural Regulation* London: Sage: 117–62

—— (1999) *Globalization and Culture* Cambridge: Polity Press

Tulloch, J. (1999) 'Fear of Crime and the Media: Sociocultural Theories of Risk' in D. Lupton (ed.) *Risk and Sociocultural Theory: New Directions and Perspectives* Cambridge: Cambridge University Press: 34–58

—— and Lupton, D. (2001) 'Risk, the Mass Media and Personal Biography: Revisiting Beck's "Knowledge, Media and Information Society"', *European Journal of Cultural Studies*, 4 (1): 5–27

Turner, B. (1991) *Religion and Social Theory* London: Sage

UNEP Report (2002) *State of the Environment and Policy Retrospective, 1972–2002* United Nations Environment Programme

Ungar, S. (1998) 'Hot Crises and Media Reassurance: A Comparison of Emerging Diseases and Ebola Zaire', *British Journal of Sociology*, 49: 1

van Loon, J. (2000a) 'Virtual Risks in an Age of Cybernetic Reproduction' in B. Adam, U. Beck and J. van Loon (eds) *The Risk Society and Beyond* London: Sage: 165–82

—— (2000b) 'Mediating the Risks of Virtual Environments' in S. Allan, B. Adam and C. Carter (eds) *Environmental Risks and the Media* London: Routledge: 229–40

Vera-Sanso, P. (2000) 'Risk-Talk: The Politics of Risk and its Representation' in P. Caplan (ed.) *Risk Revisited* London: Pluto Press: 108–32

Wales, C. and Mythen, G. (2002) 'Risky Discourses: The Politics of GM Foods', *Environmental Politics*, 11 (2): 121–44

Walklate, S. (1997) 'Risk and Criminal Victimization: A Modernist Dilemma?', *British Journal of Criminology*, 37 (1): 35–45

Waters, M. (1995) *Key Ideas: Globalization* London: Routledge

Weaver, C.K., Carter, C. and Stanko, E. (2000) 'The Female Body at Risk: Media, Sexual Violence and the Gendering of Public Environments' in S. Allan, B. Adam and C. Carter (eds) *Environmental Risks and the Media* London: Routledge: 171–83

Weber, M. (1930) *The Protestant Ethic and the Spirit of Capitalism* London: Allen and Unwin

Weinstein, N.D. (1987) 'Unrealistic Optimism About Susceptibility to Health Problems: Conclusions From a Community Wide Sample', *Journal of Behavioural Medicine*, 10: 481–95

Wilkins, L. and Patterson, P. (1987) 'Risk Analysis and the Construction of News', *The Journal of Communication*, 37: 80–92

Wilkinson, I. (2001) *Anxiety in a Risk Society* London: Routledge

Williams, C. (1998) *Environmental Victims: New Risks, New Justice* London: Earthscan

Wilson, K. (2000) 'Communicating Climate Change Through the Media: Predictions, Politics and Perceptions of Risk' in S. Allan, B. Adam and C. Carter (eds) *Environmental Risks and the Media* London: Routledge: 201–17

Winchester, S. (2003) *Krakatoa: The Day the World Exploded* London: Viking

Woodward, K. (1997) 'Concepts of Identity and Difference', in K. Woodward (ed.) *Identity and Difference* London: Sage: 8–59

—— and Watt, S. (2000) 'Knowledge in Medicine and Science' in D. Goldblatt (ed.) *Knowledge and the Social Sciences* London: Routledge: 8–40

Woollacott, M. (1997) 'The Future is Behind Us', *Guardian*, 26 February: 17

—— (1998) 'Risky Business, Safety' in J. Franklin (ed.) *The Politics of Risk Society* London: Polity Press: 47–9

Wylie, I. (1998) 'Mad Cows and Englishmen' in S. Ratzan (ed.) *The Mad Cow Crisis: Health and the Public Good* London: UCL Press: 69–73

Wynne, B. (1989) 'Frameworks of Rationality in Risk Management: Towards the Testing of Naive Sociology' in J. Brown (ed.) *Environmental Threats: Perception, Analysis and Management* London: Belhaven: 33–47

Wynne, B. (1992) 'Misunderstood Misunderstandings: Social Identities and Public Uptake of Science', *Public Understandings of Science*, 1: 281–304

—— (1996) 'May the Sheep Safely Graze? A Reflexive View of the Expert–Lay Knowledge Divide' in S. Lash, B. Szerszynski and B. Wynne (eds) *Risk, Environment and Modernity: Towards a New Ecology* London: Sage: 44–83

Index